Edward Ellis

A practical manual of the diseases of children, with a formulary

Third Edition

Edward Ellis

A practical manual of the diseases of children, with a formulary
Third Edition

ISBN/EAN: 9783337215330

Printed in Europe, USA, Canada, Australia, Japan

Cover: Foto ©berggeist007 / pixelio.de

More available books at **www.hansebooks.com**

OF THE

DISEASES OF CHILDREN

WITH A

FORMULARY

BY

EDWARD ELLIS, M.D.

LATE SENIOR PHYSICIAN TO THE VICTORIA HOSPITAL FOR SICK CHILDREN; LATE PHYSICIAN
TO THE SAMARITAN HOSPITAL FOR WOMEN AND CHILDREN; AND FORMERLY OBSTETRIC
PHYSICIAN'S ASSISTANT TO UNIVERSITY COLLEGE HOSPITAL

THIRD EDITION

NEW YORK
WILLIAM WOOD & COMPANY
27 GREAT JONES STREET
1879

TO

SIR WILLIAM JENNER, Bart., M.D., K.C.B.
D.C.L., F.R.S.

PHYSICIAN IN ORDINARY TO HER MAJESTY THE QUEEN,
AND TO HIS ROYAL HIGHNESS THE PRINCE OF WALES; PHYSICIAN TO UNIVERSITY
COLLEGE HOSPITAL, ETC.

AS A SLIGHT BUT SINCERE TOKEN OF

ADMIRATION, GRATITUDE, AND REGARD

This Work is Dedicated

BY HIS FRIEND AND FORMER PUPIL

THE AUTHOR.

PREFACE TO THIRD EDITION.

"How to detect disease is a thoroughly worked problem, but how to cure disease is one that has received too little attention from scientific physicians." The truth of this remark, which occurs in the "British and Foreign Medico-Chirurgical Review" for January, 1869, impresses itself more forcibly upon me every year that I practise my profession. In revising a book like the present, where one is necessarily carried over a vast amount of medical literature, European and American, one cannot fail to be struck with the changes that have occurred in the text-books say of the last ten years.

Bleeding, blistering, setons and issues are fast vanishing from view, but I cannot say that treatment generally has yet assumed that definite and agreed character to the attainment of which, however, at no very distant date, we may nevertheless hopefully look forward, inasmuch as the study of therapeutics is attracting more and more attention at home and abroad.

Feeling these matters strongly, I have aimed throughout in the revision of this book to render the treatment recommended as thoroughly modern, practical, and trustworthy as possible. With regard to some new remedies and methods referred to, I have had good opportunities of testing their practical value, and can therefore speak with some confidence respecting them.

Conducting this work of revision at so great a distance from home, I have to express my obligations to my friend Dr. Grimes, of Aigburth, Liverpool, for his kindness in undertaking the correcting of these sheets during their passage through the press. I am, however, alone responsible for the opinions and statements in the book.

The present edition contains much fresh matter in almost every section, and I have also added a brief dietary. I therefore trust that in spite of many shortcomings it may not be found less useful to the student and junior practitioner in England, America, and Australasia, than its predecessors.

AUCKLAND, NEW ZEALAND,
January, 1878.

PREFACE TO FIRST EDITION.

The object I have endeavored to attain in the preparation of this Manual is to present concise, yet thoroughly practical descriptions of the principal diseases of children.

I have omitted, as tending to expand too far the purpose of this book, all clinical records of my own or of others, and all controversial questions. For the same reason I have disregarded theories still under discussion. Hence cholagogues, alteratives, and counter-irritants are spoken of, because the profession attaches to these names qualities which are true qualities, though old theories of their actions may be incorrect. I have consulted the works of many distinguished authorities, both English and foreign; carefully comparing them with the results of my own experience, and to all to whom reference is made in the course of the work I desire cordially to express my acknowledgments. I also take this opportunity of thanking my friend and colleague, Dr. Hayward, for many valuable suggestions while these sheets were passing through the press.

Considerable care has been bestowed to render the Formulary which is appended useful, and as my aim throughout is essentially practical, I venture to hope that this little volume, notwithstanding many imperfections, may prove of value to students and practitioners, as a handy book of reference upon a class of diseases presenting peculiar difficulties, and yet of the deepest interest because so continually met with in practice.

118 WARWICK STREET,
   ECCLESTON SQUARE,
      September, 1869.

CONTENTS.

CHAPTER I.

	PAGE
GENERAL OBSERVATIONS ON MANAGEMENT AND DIET...................	1
1. Medical Examination.....................................	1
2. General Management during the First Year of Life...............	5
3. Diet Table for Children One Year old and upwards...............	12

CHAPTER II

GENERAL DISEASES..	14
1. Scrofulosis...	14
2. Tuberculosis...	16
3. Rachitis—Rickets.....................................	18
4. Syphilis...	21
5. Acute Rheumatism...................................	24

CHAPTER III.

SKIN DISEASES...	27
1. Exanthemata..	27
Roseola...	27
Erythema..	27
Urticaria...	28
Erysipelas.......................................	28
2. Vesiculæ...	29
Sudamina.......................................	29
Miliaria...	29
Eczema...	29
Herpes..	30
3. Bullæ...	31
Pemphigus, Pompholyx............................	31
Rupia...	32
4. Pustulæ...	32
Impetigo..	32
Ecthyma..	33
5. Papulæ...	33
Lichen..	33
Prurigo...	34

	PAGE
6. Squamæ	34
Psoriasis	34
Lepra	34
Pityriasis	35
7. Tubercula	35
Acne	35
Molluscum	35
Lupus	35
8. Xerodermata	35
Icthyosis	35
9. Parasitici	36
Tinea	36
Chloasma	37
Scabies	37

CHAPTER IV.

CONGENITAL AFFECTIONS AND DISEASES OF THE NEW-BORN	39
1. Diseases of the Navel	39
2. Sclerema	40
3. Ophthalmia (Neonatorum)	41

CHAPTER V.

FEVERS	43
1. Measles—Rubeola—Morbilli	43
2. Rötheln (Rubeola notha)	45
3. Scarlet Fever—Scarlatina	45
4. Typhoid Fever	51
5. Typhus Fever	56
6. Intermittent Fever—Ague	58
7. Variola (Smallpox)	59
8. Vaccinia—Cow-pox—Vaccination	63
9. Varicella—Chicken-pox	64

CHAPTER VI.

DISEASES OF THE BRAIN AND NERVOUS SYSTEM	66
1. Idiocy and Mental Disorder	66
2. Convulsions	67
3. Spastic Rigidity—Contraction with Rigidity	70
4. Night Terrors	71
5. Congestion of the Brain	71
6. Passive Congestion	73
7. Hæmorrhage	73
8. Tubercular Meningitis (Acute Hydrocephalus)	74
9. Hydrocephaloid Disease	79
10. Simple Encephalitis	80
11. Hypertrophy of the Brain	81
12. Chronic Hydrocephalus	82
13. Tubercle in the Brain	84
14. Chorea, or St. Vitus's Dance	86
15. Eclampsia Nutans, or Salaam Convulsion	89
16. Epilepsy	89
17. Paralysis	90

	PAGE
18. Otorrhœa	92
19. Cephalhæmatoma	93
20. Diseases of the Spinal Cord	93
21. Trismus	94
22. Hydrorachis, or Spina bifida	95

CHAPTER VII.

DISEASES OF THE AIR-PASSAGES AND THORACIC ORGANS ... 97
1. Coryza ... 97
2. Diphtheria ... 98
3. Croup (Cynanche Trachealis) ... 102
4. Tracheotomy ... 106
5. Spasmodic Croup—Spurious Croup ... 107
6. Acute Simple Laryngitis ... 109
7. Chronic Laryngitis ... 109
8. Bronchitis, or Bronchial Catarrh ... 110
9. Pertussis—Hooping-cough ... 113
10. Laryngismus Stridulus—Child Crowing ... 116
11. Pneumonia ... 117
12. Atelectasis Pulmonum ... 121
13. Pleurisy ... 122
14. Phthisis ... 125
15. Cyanosis, or Blue Disease ... 128
16. Pericarditis, Carditis, and Endocarditis ... 129
17. Epistaxis ... 133

CHAPTER VIII.

DISEASES OF THE FOOD-PASSAGES AND ABDOMINAL ORGANS ... 134
1. The Thrush ... 134
2. Stomatitis, or Inflammation of the Mouth ... 135
3. Catarrhal Pharyngitis—Sore Throat—Ulcerations in the Throat ... 137
4. Cynanche Parotidea (Mumps) ... 138
5. Tonsillitis, Quinsy, or Inflamed Sore Throat ... 138
6. Retropharyngeal Abscess ... 140
7. Dyspepsia ... 140
8. Gastritis ... 142
9. Enteralgia (Colic), Enteritis ... 143
10. Diarrhœa ... 144
11. Dysentery, or Inflammatory Diarrhœa ... 148
12. Worms ... 149
13. Jaundice ... 152
14. Acute Peritonitis ... 152
15. Tabes Mesenterica ... 154
16. Abdominal Tumors ... 155
17. Diseases of the Kidneys ... 157
18. Incontinence of Urine ... 158
19. Vaginitis ... 160
20. Prolapsus Ani ... 160

CHAPTER IX.

GENERAL THERAPEUTICAL HINTS AND FORMULARY ... 161
FORMULARY ... 162
1. Blood-restorers ... 162
2. Antacids ... 164
3. Astringents ... 167
4. Acids ... 168
5. Alteratives ... 169

		PAGE
6.	Stimulants to the Spinal Cord	172
7.	Sedatives to the Spinal Cord	174
8.	Antispasmodics	175
9.	Stimulants	175
10.	Sedatives to the Brain and General Sedatives	178
11.	Vascular and Heart Sedatives	180
12.	Nervine Tonics	183
13.	Stomachics	186
14.	Emetics	187
15.	Laxatives	188
16.	Purgatives	189
17.	Diuretics	191
18.	Diaphoretics	192
19.	Expectorants	193

External Applications—

20.	Baths	194
21.	Counter-irritants	196
22.	Gargles, Throat Applications, and Inhalations	196
23.	Liniments and Lotions	198
24.	Collyria	200
25.	Ear Lotions	201
26.	Ointments	201
27.	Hypodermic Injections	202

CHAPTER X.

DIETARY ... 204

1.	Good Nutritious Beef Tea	204
2.	Essence of Beef	204
3.	Beef Tea in Haste	204
4.	Beef and Chicken Broth	205
5.	Raw Meat	205
6.	Liebig's Food for Infants	205
7.	Chicken, Veal, and Mutton Broths	205
8.	Milk and Gelatine or Isinglass	205
9.	Milk and Suet	205
10.	Milk and Lime Water	206
11.	Bread Jelly	206
12.	Rice Cream	206
13.	Rice Milk	206
14.	Rice Water	206
15.	Barley Water	206
16.	Rice Gruel (for diarrhœa)	207
17.	Lemonade	207
18.	Refreshing Drinks	207
19.	Tamarind Whey	207
20.	Orgeat	207
21.	Egg Soup	207
22.	Rose Tea	207
23.	Jelly Water	208
24.	Iceland Moss Jelly	208
25.	Isinglass Jelly	208
26.	Chicken Jelly	208
27.	Arrowroot Wine Jelly	208
28.	Iceland Moss and Irish Moss Jellies	209
29.	Palatable Castor Oil	209
30.	Nutritious Enemata	209
31.	Stimulants	209

A PRACTICAL MANUAL

OF THE

DISEASES OF CHILDREN.

CHAPTER I.

GENERAL OBSERVATIONS ON MANAGEMENT AND DIET.

1. MEDICAL EXAMINATION.

It is very important in the study of the diseases of children to gather as much information as possible before making any manual or physical examination. The eye of the physician should inform him of a great deal, which the ear and the touch may afterwards confirm. This is a matter of the greater importance inasmuch as if we had to depend entirely upon physical examination, we might be often completely thwarted by the restlessness and excitement which are so frequently present. It is eminently fortunate if at the time of a visit the child should be asleep, as the pulse can then be felt, the breathing noted, the posture, the state of the skin, and other matters, before the child is aroused for us to see the tongue or to sound the chest. Supposing then that the little patient is asleep, what are the points that we should take note of before arousing it? First the attitude; the posture, if easy and natural, or otherwise; the color of the face, if flushed or pale; the color of the lips, if white or livid; the state of the skin, if dry or moist; if the moisture be general or restricted to the head and forehead; the expression, if natural or painful. We should note the presence or absence of moaning, starting, twitching, grinding of the teeth; the action of the nostrils, if quiet or working strongly; the eyes, if closed, or partly closed, or staring, or clenched; the respirations should be counted; the condition of the fontanelle should be considered, if closed or open, if pulsating greatly, if distended or retracted. The pulse should be noted; under two years of age it should range from 90 to 130, or in rare cases 140. Consistently with health, after three years it is rarely above 100. It may not be more than 70, yet still healthy. The actual number of beats is therefore of little value, but an infrequent pulse is a grave and valuable sign; for example, a child has had feverishness and sickness, and the pulse is found to be 130: the cause is very likely plumcake or jam; but if a child has had feverishness and sickness, and the pulse is found to be 40, the disease will most likely prove tubercular meningitis.

The size and shape of the head should be noted, if large, if the veins are full, if hot. Let the child then be aroused; its aspects should be observed waking. Does it frown or smile? is it peevish or languid? excited or resigned? Is there a dark ring round the eyes? Also the color of the face, the shape of the face, and the presence or absence of snuffling.

A perfectly healthy infant should sleep twenty out of the twenty-four hours. It is very useful in fixing the period of the commencement of an illness to inquire when the child first had broken rest or active insomnia.

Next let the child be stripped; the surface should be mottled, the flesh firm, the skin smooth and elastic to the touch, not flabby; the arms and legs should move freely. The joints should be noted, if swollen, if large or small. The respirations from one to three years of age should be 24 to 36 a minute, and diaphragmatic in character; in ordinary breathing there should be no recession of the chest-walls; this occurs in sobbing, or, if a mechanical impediment exists to the entrance of air into the lungs.

Respiration.—The number of respirations per minute ranges from 30 to 50; in early infancy 39 is the actual average.

From 2 months to 2 years, the average is 35.
" 2 years   6   "   18 during sleep, 23 awake.
" 6   "   12   "   18   "   23   "
" 12   "   15   "   18   "   20   "

Under one year the respirations vary from 40 to 50 in a minute.

Eruptions should be looked for particularly around the anus. In infants the stools should be yellow, and three or four in a day. To see the tongue advantage should be taken of the child's crying, or if it be good its lips may be touched with the fingers, and it will protrude it. Of teeth, it should have the first incisors by the seventh month, the first molars by the twelfth month, the canines by the eighteenth month, and the second molars by the twentieth month. The gums should be felt, if hot or swollen, or the reverse. Lastly, a child does not raise its head from the pillow till about the second month, and it cannot sit erect till the fourth or fifth month.

Auscultation should be practised before percussion, and the back of the chest is the most important part to auscultate in the sick child. If this be free from indications of pneumonia, bronchitis, &c., it is eminently unlikely that the front of the chest will present any. Of course when practicable both back and front of the thorax should be examined.

Dr. Vogel gives a valuable caution, viz. that dulness on the right side posteriorly is a normal physiological condition owing to abdominal pressure, whereby the abdominal organs, and notably the liver (as specially affecting the right side), are pressed upward. Dulness as indicative therefore of pneumonia, &c., should be observed in inspiration as well as in expiration, while in fact the child breathes regularly, and it should also be discernible for several consecutive days.

There are a few practical matters, in connection with the general management of children, and inferences to be drawn from peculiarities in their appearance, which, for convenience, will be briefly discussed in the present chapter. *The expression of countenance.*—1. The *upper* portion of the face is affected chiefly in brain diseases, causing knitted brow, contracted forehead, and rolling, fixed or purposeless eyes. 2. The *middle* portions of the face are changed in heart and lung affections, the nostrils are sharp or distended, or working, and there is a bluish circle round the mouth, and dark rings under the eyes. 3. The *lower* portion of the face suffers mainly in abdominal troubles, the cheeks are changed in color, sunken, puckered, the mouth drawn, the lips livid or pale—a look assumed which Sir W. Jenner describes as a Voltaire-like look. Besides these indications which are most remarkable, and of the highest practical importance, the physician will note special signs, as redness or pallor, ptosis, unequal dilatation of the pupils, and the like.

Gestures are often significant: in brain disease, the child puts its hand to its head, pulls at its hair or any covering that may be on the head, rolls the head on the pillow, and beats the air uncertainly. In abdominal disease the legs are drawn up, the face is sunken and anxious, and the child picks at the clothes. In urgent dyspnœa it tears at its throat or puts its hand in its mouth, especially when false membranes are forming, or the tongue is much furred, as in fevers, &c. Then *the cry* varies; it is labored, as if half suffocated, or better, as if a door were shut between the child and the hearer in pneumonia and capillary bronchitis; it is hoarse in croup, brassy and metallic, with crowing inspirations; in cerebral disease, especially hydrocephalus, it is sharp, shrill, and solitary, the so-called "cri hydrocéphalique," whereas in marasmus, and tubercular peritonitis, it is moaning and wailing. Obstinate and long-continued crying lasting for hours is referable usually to one of two causes, earache or hunger. A moderate and rather peevish cry attendant on suppressed cough, dry and low in character, is indicative of pneumonia. A louder shriller cry also on coughing or produced in moving the child is pleuritic. A cry accompanied with wriggling and writhing and preceding defœcation is intestinal. M. Billard distinguishes between the cry and the return, the cry proper being the respiratory act, while the return occurs during inspiration. The cry proper is sonorous and prolonged, the return is shorter and sharper; the return is feeble in young infants, but increases in strength as the child grows older. It is the return that grows weak or ceases towards the end of all diseases. Moaning is especially characteristic of disease of the alimentary canal.

Children shed no *tears* before the third or fourth month, and the saliva appears likewise about the third month. Trousseau considers that in children under two, and even up to seven years, shedding tears is a most favorable prognostic, whereas the absence of them is the reverse.

Posture.—The most characteristic posture besides the upright one of dyspnœa is the so-called "En chien de fusil," that is, on the side, the legs strongly flexed and the arms drawn close to or over the chest; this is very marked in the later stages of tubercular meningitis and some other brain affections.

The Tongue.—The following are the chief indications derivable from observation of the tongue. 1. A furred tongue with whitish curd scattered over it, indicates dyspepsia and intestinal irritation. 2. A red, dry, hot tongue points to inflammation of the mouth, stomach, &c. 3. Aphthæ often result from sheer starvation and neglect. 4. A pale flabby tongue marked at the edges with the teeth shows great debility. 5. White fur is generally indicative of fever. 6. Yellow fur of liver and stomach derangement of long standing. 7. Brown fur, of a low typhoid condition. Besides these, special conditions, as the "strawberry" tongue of scarlatina, the glazed tongue of dyspepsia, &c., will be noted under the special diseases they characterize.

Temperature.—The normal temperature of the child, taken say under the armpit, is 88° to 98° F.; if over 100° F., fever undoubtedly exists, and the cause should be searched for. No indication is more simple or more valuable than that supplied by the thermometer; by its aid alone we are often led to suspect the advent of typhoid, or scarlet fever, or to detect some latent pneumonia, or tubercle producing irritation, or worms, or some other malady which we had overlooked. It should be remembered that rigors do not occur in young children, but that convulsions and delirium correspond in a great measure to rigors and headache in the adult. Temperature is a better guide than the pulse in the diseases of young children, and should be used to correct its indications.

Dr. Finlayson has bestowed much attention on the subject of temperature in young children, and his observations go to show—

1. That there is a fall of temperature normally in the evening amounting to 1°, 2°, or even 3° Fahr.
2. This fall may take place before sleep begins.
3. It is usually greatest between 7 and 9 p.m.
4. The minimum is at or before 2 a.m.
5. After 2 a.m. it again rises, and that independently of food, &c., being taken—rises in fact during sleep.
6. The fluctuations between breakfast and tea are usually trifling.
7. The rise in a day to 104° or 105° Fahr. precludes typhus and typhoid, not scarlatina.
8. In typhoid a gradual increase for the first four days with morning remissions is diagnostic. (Wunderlich.)
9. In tubercle the evening temperature is as high, or, according to Dr. Ringer, higher than the morning.

The Eye.—Squinting in acute illness is a grave prognostic; it may occur from reflex irritation or from paralysis, or from convulsions, but the convulsions may cease and the squint remain for a while or even permanently. When strabismus occurs in tubercular meningitis it is an almost fatal sign.

A small pupil is not so common as a large one; it occurs in active congestion, in opium poisoning, and in sleep. It should be remembered that the eye is always more or less turned up beneath the upper lid. Large pupils if equal in size are only of grave import when insensible to light; inequality of the pupils coming on in acute illness is a very grave prognostic. M. Jadelot has noticed that the form of the pupil is irregular in children suffering from the intestinal irritation of worms.

The Pulse varies from 110 to 150 consistently with health; it may be irregular consistently with health. It is rather quicker in the female than male after seven years; it is somewhat slower during sleep. A very slow pulse is an indication of cerebral disease.

Table of the Pulse. (*Müller.*)

At birth	130 to 140
1st year	115 " 130
2d "	100 " 115
3d "	90 " 100
7th "	85 " 90
14th "	80 " 85

The following aphorisms of Bouchut are of the highest practical value:

1. In early childhood there is no relation between the intensity of the symptoms and the material lesion. The most intense fever with restless-

ness, cries and spasmodic movements, may disappear in twenty-four hours without leaving any traces.

2. Abundant perspiration is not observed in very young children; it is entirely replaced by moisture.

3. Fever always presents considerable remissions in the acute diseases of young children.

4. In the chronic diseases of infancy fever is almost always intermittent.

5. When children are asleep their pulse diminishes from 15 to 20 pulsations. The muscular movements which accompany cough, crying, agitation, &c., raise the pulse 15, 30, or even 40 pulsations.

6. The diseases of youth always accelerate the process of growth.

A child grows most rapidly in the first weeks of life, e.g. in the first year it should grow from 6 to 7 inches. From the 4th to the 16th year about 2 inches yearly; from the 16th to the 17th year $1\frac{1}{2}$ inches; from the 17th to the 20th year, 1 inch. Disease of the bones, rickets and scrofula, retard growth.

A child should run alone at the end of a twelvemonth, and if when it has commenced to walk it uses chiefly its toes, and has a limping gait, more especially if pain be complained of in one knee, and tenderness be caused by handling the limb, commencing hip-joint disease may be inferred.

2. General Management during the First Year of Life.

The average weight of a newborn child is 7 lbs.; the extremes are from 4 to 11 lbs. The following are the chief anatomical peculiarities of the newborn infant and the practical inferences deducible therefrom. A newborn child has a small stomach, therefore do not suffer it to be overstuffed with food. Its intestinal action is more rapid, and the power of generating heat is small, hence the great need of warmth. Its heart is large, its brain is large, and contains less relative phosphorus than the adult brain. The brain of an idiot contains the least phosphorus of all, a possible hint for treatment. It is worth remembering that, owing to the large size of the liver in infants, a child is often sick when placed on its left side after suckling, owing to compression of the stomach by the weight of the liver. So if difficulty occur in a child sucking the right breast, its legs should be turned under its mother's right arm, so that it may be allowed to suck lying on its right side.

A child should be washed in warmish water twice a day, carefully dried, and dusted over with unscented starch-powder; if there be any excoriations oxide of zinc will replace the starch with advantage. An infant's clothes and napkins should never be washed in soda. If on the separation of the navel there is a little ulceration and serous exudation, zinc ointment is the best remedy, or a little bismuth ointment. If the navel be ruptured, it should be immediately secured with soap-plaster or diachylon, a small pad of lint over a piece of flat cork being placed on the protrusion; an elastic bandage over all is useful to keep things in place. In severe rupture a piece of sheet lead should be folded over the cork and well padded with lint, and then fastened as before.

The child's ordinary flannel bandage should not be left off till the third month, when it may be dispensed with gradually by using smaller ones. If pertussis or any straining cough be present, the band should not be left off even then.

No mother should willingly delegate to another the duty of nourishing her own offspring; no food whatever agrees with a child like its mother's milk. "A mother who does not suckle is more liable to peritonitis, inflammation of the uterus, abscesses, cancer of the breast and womb" (Decaisne). At the same time there are cases in which suckling is evidently beyond the mother's strength, or it may be inadvisable by reason of some hereditary taint, as phthisis, syphilis, &c. Otherwise the child should be put to the breast within a few hours of birth; the uterus is thereby induced to contract more thoroughly, and the child gains the benefit of the "colostrum," or earliest milk, which saves all necessity for castor-oil, honey and butter, and other recipes of old nurses. If the child be tongue-tied, the frænum linguæ will require snipping with a pair of scissors. Mr. Maunder recommends that this little operation be done with the nail. The child should be applied alternately to each breast at intervals at first of an hour and a half to two hours, gradually increased to three and four hours as the child grows older. As a rule no artificial food whatever should be permitted when the breast of milk is good; at any rate until the sixth or seventh month.

The best test of the quality of the mother's milk is the fact that the child thrives or does not thrive. If there be any doubt as to the quantity of the supply, the infant may be weighed before and after suckling; the increase in weight should not be less than three ounces under three months, and about six ounces in an older child. If the catamenia come on during nursing, or if conception occur (which is a rare event), the child will often pine and lose flesh. This seems to be, indeed, one of the predisposing causes of rickets. Cazeux suggests that phosphorus should be given for a time in some form under such circumstances, while the weaning, which it will be often necessary to effect soon, is being accomplished.

In cases where artificial food is absolutely necessary, cow's milk and water and sugar, one-third milk and two-thirds water ; or, better still, lime-water (to prevent the curdling of the cow's milk in the stomach) with a little loaf sugar or sugar of milk may be given. When this disagrees, as it sometimes will, I have occasionally ordered the sugar in any form to be omitted, and a few grains of ordinary table salt to be substituted, with very satisfactory result. Brown sugar is never to be given. As the child grows older less water, or lime-water, and more milk, may be given, until after five or six months the milk may be given alone.

Dr. Hiram Corson condemns the practice of giving the milk diluted. He affirms that it should be given pure, and that much of the early mortality amongst children means simply starvation from this cause. But Dr. Corson appears to lose sight of the fact that the essential and vital difference between cow's and woman's milk consists, not so much in the relative quantities of this or that solid constituent, as in the fact that the casein of woman's milk always coagulates in small floccular pieces—broken flakes—while cow's milk (and other milks to a less degree) coagulates into hard, gelatinous, indigestible lumps. It is to render the cow's milk more digestible that the addition of water, and especially of lime-water, is recommended by physicians. No doubt so far as London is concerned the addition of water to milk is often a work of supererogation, but for a child with a delicate stomach the addition of a little alkali to prevent this curdling is, I am persuaded, of the utmost service.

A teaspoonful of cream may be added to a quarter of a pint of milk, as recommended by Sir W. Jenner; four or five ounces of the "food" thus prepared is amply sufficient for a meal; if the child is allowed to suck in more it will be sick from repletion. Other forms of diet adapted for occasional use or at weaning may next be mentioned, e.g., the crumb of bread boiled and sweetened with a little boiled milk added, is a good food, so are Robb's biscuits, so is baked flour, or baked flour and oatmeal (one part of oatmeal to two parts of flour); two tablespoonfuls with half a pint of milk well boiled and sweetened is a capital food. Revalenta Arabica, or lentil food, is strongly recommended by Dr. Routh. Hard's Food, Neave's Food, Ridge's Food, Soojie, Wheat Phosphates, Tapioca, and Semolina, Brown and Polson's Corn Flour, have all their advocates.

It is not at all easy to predicate which of these foods will prove most useful in individual cases. "Hard's" is apt to exercise a rather costive effect, while "Neave's" is far less binding in its operation. I have found "Ridge's" very valuable in many cases, but equally unsuitable in others. Robinson's patent groats are often useful for a few days when much constipation exists, and they may be mixed in proportion of one-half or one-third part to either of the other foods if desirable. Arrowroot made with new milk is a very nourishing food. Rice milk and rice jelly are especially useful when the bowels are disordered. Regarding the employment of the condensed milk, Dr. Pavy, in a note on p. 191 of his "Treatise on Food," refers to some remarks of correspondents of the "Lancet" of November, 1872, in which it is admitted that infants "readily take it, grow plump, and appear to thrive," but that this is in reality a delusive appearance, that such children really lack vigor, and fall an easy prey to diarrhœa and other affections. Dr. Pavy adds that the evidence adduced at present can be only looked upon as suggestive, but that the matter is important and waits further investigation, &c. My own experience is decidedly favorable to condensed milk. The "Anglo-Swiss" and the "Aylesbury" are the kinds to the value of which I can bear testimony. In the course of a long sea voyage in which a number of babies happened to be on board, and no cow, the value of the condensed milk became sufficiently apparent. Two very delicate infants simply lived upon it, the mothers having nothing for them. I happened to have a few dozen tins for the use of my family if found needful, and I was, therefore, fortunately able to keep those infants supplied when the ship ran short, but for which they must certainly have died. At the time the mothers' milk failed these children looked half starved, but they were hearty enough after nearly three months of Swiss milk diet. Further, I have known in New Zealand many mothers obtain Swiss milk in preference to all other foods, and that when fresh milk was abundant enough. I must say I heartily endorse Dr. Chambers's witty condemnation of Liebig's food for infants. "Sensible people," says Dr. Chambers, "will be content to leave the recipe for some coming race who may prefer art to nature" (see Formulæ). Artificial food of whatever kind should, as a rule, not be required oftener than twice in the twenty-four hours, from the fourth to the seventh or eighth month, while suckling. Towards the ninth and tenth months, when weaning should be commenced (to be certainly completed within the twelvemonth if not sooner), more frequent meals of artificial food may be allowed.

It is certain that some children will digest and thrive upon varieties of food which are simply deleterious to others; the powers of digestion are remarkably stronger, and the explanation of this is in my opinion to be found in the varieties of diathesis which even thus early assert their presence and demand and require diligent attention. It is the children of the rickety and of the rheumatic diatheses that suffer most severely from dyspepsia and "weak stomachs." Tubercular children are far hardier in this respect, while the strumous and the syphilitic occupy a median position. Therefore by attention to the special diathesis of a given case, more valuable hints will be afforded in the pining and wasting which so often result not merely from improper food being given, but also from the want of power to assimilate a *quantity* of food otherwise suitable enough in character. It is the child of gouty and rheumatic patients that suffers so often from acidity, whose milk is constantly returned in lumps of curd, or in whose motions such lumps are easily discoverable, and who requires most especially that his food should be alkalinized by lime-water, or some other suitable means, *e.g.* a grain of carbonate of potash to about an ounce of food. The use of the farinacea must be always guarded under the age of three months. The deficiency of saliva before that time renders the proper digestion of starchy foods difficult, and if given in any quantity they are sure to disagree.

Dr. Gumprecht recommends carrot pap for young children. One ounce of finely scraped full-grown carrot should be mixed with two cupfuls of cold water and allowed to stand for twelve hours. The residue is to be pressed and the liquor strained off. This liquor thus obtained is to be mixed with powdered biscuit farina to make a pap and then placed over a slow fire—not allowed to boil lest the albumen be coagulated. It may be sweetened with loaf sugar. This food is not suitable during any tendency to diarrhœa. Dr. Meigs of Philadelphia recommends a food thus prepared: a scruple of gelatine is soaked for a short time in cold water and then boiled ten or fifteen minutes in half a pint of water till it is dissolved. To this is added with constant stirring, and just at the termination of the boiling, the milk and arrowroot (the latter being previously mixed into a paste with a little cold water). After the addition of the milk and arrowroot and just before the removal from the fire the cream is poured in and a little loaf sugar added. The proportion of milk, cream, and arrowroot must depend on the age and digestive powers of the child. For a healthy infant within one month three to four ounces of milk, half an ounce to an ounce of cream, a teaspoonful of arrowroot and half a pint of water. The quantity of both milk and cream should be increased as the child grows older.

Constant sickness is a common trouble of mothers during the first year; the cause is almost always injudicious feeding; very often the farinacea are to blame. When the child is emaciated and the fontanelle depressed, there will be nothing like keeping it to a healthy breast of milk, or, failing that, to ass's milk, goat's milk or cow's milk and lime-water. The infant must be at the same time kept warm, especially the feet and stomach. The skin should be subjected from time to time to gentle friction and the bowels kept open. A mixture of bismuth and soda, or of hydrocyanic acid and nitrate of potash, is useful. Dr. Eustace Smith recommends—

℞ Acidi Hydrocyan. dil... ℳ vj.
    Pot. Nitrat.. ʒ j.
    Syrupi... ℨ ss.
    Aquæ, ad... ℨ iss.
Dose, ʒ j. t. d. s.

If the mother cannot suckle, a wet nurse is the best substitute. If possible she should be a healthy married young woman, without blotches or scars, especially on the neck, with regular teeth and clear complexion, whose child is about the age of the child she is to suckle, and whose child is itself healthy and free from sores, and redness about the anus. Her milk, of which a little should be examined in a glass, should be thin, bluish white, sweet, throwing up a clear cream on standing. The breasts should be of moderate size, equal, and firm (glandular not adipose tissue being required), the nipple should be of moderate size, well made, and prominent, that the child may easily seize it. A wet nurse must not be too highly fed; she should live regularly, simply, and quietly, and take daily exercise.

Amongst the wealthier classes there is no more common error than over-pampering a wet nurse; forgetting the hardships from which she has come and amongst which she has thriven and kept her health, she is straightway made to do nothing and to eat constantly of the richest food, and then such persons are surprised that the milk does not appear to agree with the child, and that the nurse looks poorly and the like.

Woman's milk should have a specific gravity of 1032·67; its analysis, according to MM. Vernois and Becquerel, is—

```
Water................................................. 889·08
Sugar..............................  ............... 43·64
Casein and extractive matters................  ... 39·24
Butter.............................................. 26·66
Salts.............................:............... 1·38
Solid constituents........................................ 110·92
                                                        -------
                                                        1000·00
```

The amount of milk secreted per diem should be from thirty to forty fluid ounces. The milk of brunettes is richer in solid constituents than that of blondes, though the latter often secrete larger quantities.

Composition of Milks (Vernois and Becquerel).

	Specific gravity.	100 parts contain—		The solid components consist of—			
		Fluid.	Solid.	Sugar.	Butter.	Casein and extractives.	Salts.
Woman......	1032·67	889·08	110·92	43·64	26·66	39·24	1·38
Cow..........	1033·38	864·06	135·94	38·03	36·12	55·15	6·64
Ass..........	1034·57	890·12	109·88	50·46	18·53	35·65	5·24
Goat.........	1033·53	844·90	155·10	36·91	56·87	55·14	6·18
Ewe..........	1040·98	832·32	167·68	39·43	54·31	69·78	7·16

Weaning should be effected at from nine to twelve months, and it should be effected gradually, that is, artificial food should replace the breast milk more and more frequently, until the breast is only given at night, and at last not given at all. Any of the foods recommended at page 7 are suitable; that which agrees best should be selected and kept to. It is not good to accomplish weaning when the child is ailing; a favorable opportunity must be sought, but it should always be accomplished by the end of the twelvemonth.

Light is of immense importance in the due development of infancy and childhood. In a remarkable paper read before the French Academy of Science, Dr. Dubrunfant points out the importance of different colored lights in animal and vegetable growths. The researches of Gratiolet, Cailletet, and others have demonstrated that the red rays of the spectrum are those to which the physiological functions exercised by the sun on plants is to be attributed. Green light, which the vegetable kingdom refuses, the animal requires. Red, the complement of green, is that which, owing to the blood, tinges the skin of the healthy human subject, just as the green of leaves is the complement of red they absorb. Hence, says Dr. Dubrunfant, red should be proscribed for our furniture, except curtains. He quotes the cases of four children who had become chlorotic by living in the streets of Paris, who regained their health without other means than exposure to the influence of the solar rays on a sandy sea coast. Whether we fully endorse all Dr. Dubrunfant's conclusions or not, they may at least serve to remind us of the great importance of light as a preventive of, and remedial agent in, disease.

Dentition commences usually at the seventh month, but, especially in rickets, it may be deferred till the eighteenth month or second year. When a child is born with teeth they usually fall out. The temporary teeth (twenty in number) are generally cut in pairs. The following table indicates in *months* the usual times of their appearance:

Molars.	Canine.	Incisors.	Canine.	Molars.
24—12	18	9 7 7 9	18	12—24

The lower jaw usually is a little in advance of the upper. The permanent teeth (thirty-two in number) appear as under in *years*.

Molars.	Bicuspids.	Canine.	Incisors.	Canine.	Bicuspids.	Molars.
25 13 6	10 9	11	8 7 7 8	11	9 10	6 13 25

At 2 years the child has altogether - - 16 teeth.
At 2½ "   "   - - 20 "
At 6* "   "   - - 48 "

During dentition the child's health requires unusual care, the bowels must be regulated, the diet strictly attended to, the gums lanced when they are hot and swollen, but not otherwise; the diarrhœa of teething is natural, and if in moderation should not be interfered with. A little castor-oil is the remedy when this diarrhœa becomes griping, offensive or troublesome. Astringents do harm nine times out of ten in such cases.

The period of dentition is most frequently a trying one, even with children otherwise healthy; it is peculiarly so in a child the subject of any of the diathetic diseases. This is the time when ulcerated mouths, disordered bowels, green stools, convulsions and congestion of the brain are especially common. Paralysis, too, occurring suddenly, without warning, the child going to bed well, and after a "restless night," the mother is alarmed to find in the morning one arm helpless, perhaps an arm and a leg; more rarely one arm and both legs, or both arms without the legs

---

* At six, 20 deciduous and 28 permanent make 48.

being at all affected. Generally this paralysis is temporary, lasting a few weeks only; occasionally it lasts for mouths, sometimes it is incurable. Dr. Fliess considers that the molar teeth are the usual promoters of dental paralysis, more rarely the incisors. It is usually recommended to lance the molars when about to protrude with a crucial incision. This may be very advantageously done in many cases; in strumous and rickety children it is of more doubtful efficacy, and may be followed by ulcerative stomatitis and occasionally by a protracted oozing of blood from the cut surfaces very troublesome to control. Some pain and difficulty in passing water is also a common concomitant of dentition, permitting a free use of demulcent bland drinks; and the use of citrate of potash, with a drop or two of Battley where there is much pain, form appropriate treatment. In convulsions during teething, hydrate of chloral in four grain doses for a child eight years old is useful. The taste of hydrate of chloral is best disguised for children by a little tincture of orange-peel and peppermint water well sweetened.

At the end of the first year it is time enough to commence more solid and general diet, then bread and gravy, mashed potato and gravy, and by-and-bye, as the child gets teeth, small pieces of meat cut fine and mixed with potato and gravy, light puddings, and general diet which will be found indicated in the following diet tables :

DIET TABLE.

Meal	Low.	Ordinary.	Extra, comprising ordinary and some of the following in certain cases.
Breakfast, 8–9 o'clock.	Bread scalded with milk, and water in equal proportions. Gruel. Arrowroot. Rice milk. Milk and lime-water (¾–¼). Pearl barley boiled in milk.	Half a pint of hot new milk poured on a slice of bread, and some bread and butter to eat with it. A little *loaf* sugar may be added to the bread and milk. Bread and butter, weak *black* tea, or better, cocoa or chocolate with plenty of milk. Chocolate is very nutritious and wholesome, and children soon like it.	Yolk of new-laid egg beaten up in tea with a teaspoonful or two of cream. Lightly boiled new-laid egg. Iceland Moss cocoa.
Dinner, 12–1 o'clock.	Bread and gruel. Light puddings, as sago, tapioca, semolina, bread, rice, tous les mois, corn flour, &c. Fish, as boiled sole, whiting. Boiled chicken. Weak veal-tea. Weak chicken broth.	Bread, mashed potato, and gravy. Beef tea, veal tea, chicken broth, and mutton broth. Light puddings, rice, custard, vermicelli, sago, tapioca, corn flour, &c. *Fish.*—Turbot, soles, whiting, smelts, flounders, fresh cod, mullet, in all cases boiled rather than fried. *Meat.*—Roast mutton, boiled mutton, mutton chop, lamb, roast beef, chicken, pigeon, rabbit, turkey. *Vegetables.*—Mashed potatoes, cauliflower, brocoli, spinach, turnips, parsnips, carrots, French beans, asparagus, sea-kale, vegetable marrow, lettuce.	Clear soups made from the lean of beef, veal, or mutton, and thickened with sago, vermicelli, maccaroni, rice, pearl barley or wholesome vegetables, and not highly seasoned. Jellies, calf's foot, Iceland and Irish moss. Liebig's Extract of Meat. Turtle soup; clear turtle is very digestible and highly nutritious. Raw meat made by shredding mutton or beef quite free from fat, and pounding till it becomes a pulp; it should then be carefully strained and about a teaspoonful given at a time. This is very valuable in protracted diarrhœa and in exhausting diseases; the quantity given may be increased if the sto-

GENERAL MANAGEMENT.

		mach retains it well. It causes offensive evacuations. Oysters, lamb's sweetbread, whitebait. Larks, pheasant, snipe. Stimulants, ordinary, sound hock or claret, bitter ale, stout; extraordinary, port wine, brandy, champagne.
	Fruit.—Most wholesome when baked or stewed; perfectly *ripe* fruit is also wholesome in moderation. *Beverage.*—Water, toast and water, sometimes milk and water.	
Tea, 4 o'clock.	As breakfast. If weak tea be taken in the morning, cocoa may be taken now or *vice versâ*.	Gelatine, isinglass, or suet tied in a muslin bag and boiled in milk, to be subsequently sweetened with white sugar. As breakfast.
Supper, 6 o'clock.	Thin gruel. Milk and water. Arrowroot.	Gruel, rice pudding, corn flour, arrow-root, as puddings or blanc mange, with a piece of bread. Light puddings. Beef tea. A little jelly or blanc-mange, with bread.

Children's meals should be regular in time, and a child should be put to bed soon after its supper; the meal before bedtime is never to be a heavy meal.

To be Avoided.—All rich and highly-seasoned soups. *Meats.*—Pork, veal, bacon, salt beef, duck, goose, sausages, liver, kidney, heart, tripe. *Fish.*—Crab, lobster, in fact, all shell-fish except oysters, and those only as occasional aliments in extra diet. Salmon, salt cod, eels, sprats, herrings, mackerel, &c. *Vegetables.*—Cucumber, radishes, celery, onions, parsley, and flavoring herbs. Pickles, pastry, sweets, sauces, spices, nuts, cheese, sweet cakes, suet pudding.

The use of the farinacea is a matter requiring a little care. In some cases they are positively injurious—as, for instance, in some diseases of the stomach and intestines; in others—as in some of the diathetic diseases—they are beneficial when the child is able to digest them. When, therefore, the bowels are loose and slimy, or there is some vomiting and other signs of dyspepsia, the quantity of farinaceous food should be diminished until, as tone is restored to the stomach by appropriate remedies, it is enabled to digest them, when they will be found useful and nutritious.

In a general way the simpler a child's diet is, whether it be well or ill, the better, but cases of dangerous and protracted illness, especially amongst the children of the rich, will often tax the ingenuity of the physician, to permit variety without unwholesomeness, or to coax a pampered appetite without indiscretion. (See Dietary.)

CHAPTER II.

GENERAL DISEASES.

1. Scrofulosis.

It is of great importance in studying the diseases of children to get a clear idea of the various diathetic states that mask and change the course of ordinary acute diseases. Scrofula is a constitutional affection of childhood; it is, in fact, more limited to childhood than tuberculosis, and it is most important that its essential features should be clearly recognized. Sir W. Jenner has drawn particular attention to many of the points that distinguish tuberculosis from scrofulosis. The latter is characterized pathologically by affecting especially — the lymphatic glands, causing strumous inflammation and abscess; the mucous membranes, *e.g.*, strumous ophthalmia; the skin, producing obstinate and very chronic cutaneous diseases; and caries of the bones.

In cutaneous diseases the formation of pus is very characteristic of the scrofulous type, hence vesicular and pustular eruptions are the rule, as contrasted with the papular or neurotic eruptions, of the tubercular diathesis. Eczema, impetigo, and ecthyma may be named as especially common.

The temperament of a child afflicted with struma is phlegmatic; the mind and body are backward; such a child is dull and heavy; its skin is thick and muddy-looking, its complexion doughy, its upper lip thick ("the strumous lip"), the nostrils wide, and the alæ of the nose thickened, the lymphatic glands, especially the cervical chain, enlarged, or becoming so on the slightest provocation. The abdomen is tumid, the ends of the bones are rather large, and the shafts thick.

Scrofula resembles tuberculosis in being hereditary, and in being readily induced by defective hygienic conditions, such as unwholesome and insufficient food, bad ventilation, scanty clothing, and the like. It also resembles tuberculosis in frequently becoming associated with phthisis and hydrocephalus.

There are a few manifestations of struma it will be convenient to notice here:—1. The formation of abscesses, which often take place very early in life in the subcutaneous areolar tissue. The essential characteristic of a strumous abscess is *indolence*; they are not tender; they increase but slowly, and they leave deep and often permanent scars.

The cervical glands are perhaps the most prone to be affected in the scrofulous diathesis; these readily enlarge in teething, in slight attacks of stomach and intestinal disorder, and especially on the occurrence of any acute disease. The mother commonly states that the glands of the ears are down, and we find the whole chain like a string of beads round the neck. Sometimes one or other of them going on to slow suppuration, the neighboring glands become similarly affected; sinuses are formed, and hence the unseemly marks and scars often seen in the necks of persons of the strumous type. Catarrhs, bronchial, gastric, and intestinal, are common accompaniments of the strumous diathesis; the child is always catching cold, and is especially prone to a watery diarrhœa, which is occasionally very intractable to treatment.

2. Otorrhœa is one of the very commonest accompaniments of the strumous diathesis. Besides the ordinary treatment of scrofula presently to be noted, the ears should be well syringed out daily with warm water or a weak zinc lotion, and a sinapism about the size of a florin should be frequently applied over the mastoid process.

3. Ozæna (rhinorrhœa) is also common; the amount of heat and swelling about the lining membrane of the nostrils is often considerable, and with the offensive mucopurulent discharge make this complaint a very distressing one. When we are satisfied by the use of the speculum that there is no polypus or other nasal tumor, the local treatment must be directed to removing the offensive crusts of mucus by frequently syringing with weak Condy's fluid lotion (℥ ij. to Oj.), or chloride of zinc (gr. xvj. to Oj.), or chlorinated soda or other disinfectant lotion, and directions must be given that the syringing be thoroughly and effectually performed, after which it is well to have the nostrils anointed with some stimulating ointment, as the Ung. Zinci or Hyd. Nit. Mitius, to diminish the amount of secretion. The stomach and bowels are frequently deranged in these cases, and should receive attention. Occasionally ozæna is a concomitant of the syphilitic diathesis; the local treatment will be the same, but the constitutional remedies will then of course require to be directed against the syphilitic taint in place of the strumous.

4. Ophthalmia.—Strumous ophthalmia accompanied with ulcers in the cornea, and often spasm of the lids is very common. The most characteristic symptom is great intolerance of light. There is also profuse lachrymation; the eyelashes grow long and wavy, and after a time fall out, to be replaced by short, fine, thinly-distributed hairs which are permanent. The disease is both extremely obstinate and also very liable to recur. Both eyes should be protected by a shade, whether one only or both be affected. Pure air, tonics, and the general hygienic and dietetic treatment of struma must be put in force. A scruple of belladonna in half an ounce of glycerine makes a good application to be smeared round the eye to relieve the intolerance of light and the pain so often present, or a solution of atropine (gr. j., aquæ ℥ ij.) may be dropped into the eye several times a day. Warm lotions, when lotions are needed, are to be preferred to cold; but all lotions and ointments while pain exists are injurious, and increase the irritation. In the later stages, however, weak lotions of Vinum Opii and nitrate of silver may be dropped into the eyes, and the weak citrine ointment smeared round the lids at bed-time. When the spasm of the eyelids is extreme Dr. Swanzy advocates von Graefe's treatment, which consists in dipping the child's face in a basin of cold water and holding it there for ten seconds; the child is then allowed to take a breath, and the process repeated several times. The effect is described as being magical, the child readily allowing its eyes to be examined, and atropine injected. The treatment may have to be repeated if spasm recur, but it is always eventually successful in subduing it. It must never be forgotten that no local treatment will be successful without constitutional treatment be also used and persevered in.

With regard to the general management of struma, a few words must first be said about prophylaxis. If struma exist in the parents, or in either of them, the mother must observe unusual care during utero-gestation; she should live by rule, wear warm clothing, avoid excitement, and take regular exercise; it is far better for a strumous mother not to suckle her child, but a wet nurse should be procured from the first. When the child is weaned extra care must be used, all improper diet must be rigorously excluded, remembering that what another child can do with impunity and perhaps with advantage, will in the strumous subject only serve to develop the taint more quickly and certainly. It will be wise to keep such a child, therefore, to good cow's milk, impregnated perhaps with mutton suet, according to the plan of Dr. Paris; to give it weak veal tea and other light broths, to let it have its potatoes mashed with milk, and to be careful in the use of vegetables; such a child should be warmly clad, not suffered to run with its bare legs exposed to the cutting east winds as so many in this city do, or to be cooped up in a perambulator with the said wind full in its face, till it is blue with cold. A strong child will live through such things (they are not good for them, but strong children survive them), strumous children will not. A bath in which a handful or two of Tidman's sea salt has been placed over night will be useful, but the child should after bathing be well rubbed with a Turkish towel to excite the glow and healthy action of the skin. Of medicines, iodide of potassium in small doses is useful, so is the syrup of the phosphate of iron, and the syrup of the iodide of iron; perhaps the best medicine is cod-liver oil and lime-water in equal parts, taken twice or three times a day. Dr. Vogel recommends a tea made from walnut leaves, of which three or four cupfuls are to be given in the day. The state of the bowels must also be carefully attended to. Should abscess occur it should be opened with a small opening, and the strength well supported during suppuration. The enlarged cervical glands, and indeed strumous enlargements anywhere, often improve wonderfully on being painted with iodine paint, or even with the ordinary tincture of iodine. It is of great importance to remember that mercury in all its forms is borne very badly by scrofulous children. I have repeatedly seen them get much worse under its use. Lastly, the mineral waters of Heilbronner, Kreuznach, and in England of Woodhall Spa, are often serviceable.

2. Tuberculosis.

In this most important and common diathetic condition the child has a nervous system highly developed, the mind and body are alike active, the figure is slim, little adipose tissue being made; the organization delicate and refined. Sir W. Jenner thus sums up the chief characters, "Thin skin, clear complexion, the surface veins distinct, eyes bright, pupils large, eyelashes long, hair silken, face oval, ends of bones small, shafts thin, limbs straight." Dr. Gee thinks freckles a sign of much value in the diagnosis of tuberculosis.

Children subjects of tuberculosis cut their teeth early, run alone, and talk early. The leading pathological tendencies are fatty degeneration of liver and kidneys, deposition and growth of tubercle, inflammation of serous membranes.

The thorax is generally long, almost circular. It may be long and also pigeon-breasted, which is distinguishable from the pigeon-breast of rickets by being limited to the lower part of the chest; the upper being somewhat flattened; in rickets the deformity extends as high as the second rib

(Jenner). The cause of the long pigeon-breasted thorax is repeated attacks of catarrh and bronchitis.

Tuberculosis manifests itself frequently in phthisis, in hydrocephalus, and in so-called tabes mesenterica, but tuberculosis in the child often exists without affecting the lungs; thus, for instance, the bronchial glands may be attacked, the lungs remaining free.

Tuberculosis is undoubtedly hereditary, and, like scrofulosis, favored in its production by improper and scanty food, ill-ventilated rooms, and unhygienic conditions generally. It is important to remember that tuberculosis never produces scrofula nor scrofulosis tubercle, the two diathetic conditions being distinct and separate; for though in the course of scrofula signs of tubercle may be developed, the contrary does not occur, and the two types oftener remain completely distinct. Tuberculosis may be acute or chronic. When acute the pyrexia is considerable, the emaciation rapid, and death may occur in a few weeks; or the disease may lapse into a chronic condition, evidenced by great anæmia, listlessness, want of firmness of the flesh, and a subpyrexial condition, in which thirst is a prominent symptom, and there is considerable heat of skin, some irregularity of the bowels, a little cough, vague and fleeting pains, and increasing debility; a temperature frequently over 100° F., especially towards night, is a constant symptom, and is a valuable aid in diagnosis. At the same time the temperature may be very variable, e. g., 104° F. or 105° F. at night, and yet fall within a few hours to 98° or 99°. Perspirations are not uncommon, especially towards morning, but they are general, and not confined chiefly to the head, as in rickets. A peculiar, almost harsh dryness of the palms of the hand and soles of the feet is a very common symptom.

With regard to treatment, as in scrofulosis, it is all-important that prophylactic precautions should be carefully adopted. The child of tubercular parents must have a healthy wet nurse provided; all hygienic conditions of feeding, exercise and sleeping, scrupulously observed; above all, fresh air in abundance will be of the utmost service. The mental powers are not to be excited in any way; it is well to allow such children to remain "backward;" on the other hand, exercise of suitable character is to be daily permitted, out of doors when possible (and that is always, except in damp weather and during prevalence of east winds); in-doors in some hall or spacious room, where romping can be indulged in. Such children will often bear heat well or cold well, but they will suffer acutely from sudden transitions; hence such should be guarded against. Damp is probably the busiest factor of tuberculosis; hence a marshy, ill-drained neighborhood is on all accounts to be avoided. If the child is to winter abroad, a dry warm climate or a dry cold climate may be selected, but a moist warm or cold damp climate will be most prejudicial. Daily salt baths are useful, and friction should always be employed after them over all the skin. For diet, milk, cocoa, chocolate, the yolks of new-laid eggs beaten up, cream, white fish, chicken, mutton, and clear but not rich soups are useful. The farinacea are to be used with moderation; the digestion is sometimes not strong enough for such food, and, if undigested, the bowels are only irritated, and harm rather than good results. For the diarrhœa which is so easily excited in these patients nothing is so good as phosphate of lime, half a grain to a grain or two grains several times a day. For general medicines, cod-liver oil, glycerine, the syrups of the iodide and phosphate of iron, are the best. Citrate of potash is highly spoken of by Dr. Buchanan. One tonic at a time is enough, and it is good practice to change it after a time, or even to omit it altogether for a few weeks. Slight ailments, unworthy of notice in healthy children, must receive attention in the subjects of tuberculosis, e. g., slight stomach and bowel disorders, headaches, unusual fatigue, and the like. By such judicious management, carried on year after year with unremitting patience, I have seen children whose lives have been despaired of live to grow up, if not strong, yet still sufficiently so for all the ordinary duties of life.

5. Rachitis—Rickets.

A constitutional disease, characterized by general cachexia, a peculiar condition of the bones, and often by albuminoid degeneration of some portion of the glandular system. Rickets is essentially a disease of childhood. It is rarely congenital, and probably not hereditary, although the children of drunken, syphilitic, and scrofulous parents are the most prone to develop it. So unhygienic conditions, bad ventilation, scanty and improper food, insufficient clothings, are frequent factors in its production. Of these, perhaps, improper food occupies the most conspicuous place. Rickets is, unfortunately, a very common disease; in fact, Sir. W. Jenner says of it that it is "the most common, the most important, and in its effects the most fatal, of diseases which exclusively affect children."

It is fair to say that rickets does not in all respects fulfil the condition of a diathetic disease. "Reduce the strength," says Dr. Eustace Smith, "to a given point and rickets begins. . . . Rickets is a disease which must run its course; by judicious treatment it may be stayed at any point of its career. . . . It is acquired under the influence of certain causes, lasts as long as those causes continue in operation, and unless the structural changes are so extensive and the general health so reduced as to forbid recovery, passes off when the causes are removed."

The period at which rickets is at first usually manifested is about that of the first dentition, though it may be developed as early as the fourth month. The precursory symptoms are such as are common to many diathetic diseases, namely, fretfulness, irritability of temper, capricious appetite, disordered state of the bowels, with offensive leaden-colored stools, thirst, fulness of the abdomen (partly in consequence of the diminished capacity of the thorax and pelvis, and partly in consequence of the weakness of the abdominal muscles and the flatulent distention of the intestines and the enlargement of the liver and spleen—the abdomen is almost always enlarged in rickety children), emaciation, slight pyrexia, some tenderness and swelling of the joints, especially of the wrists and ankles, and the bones are thickened just outside the sutures, the superficial veins enlarge, the fontanelle remains open.

Dr. Vogel mentions a condition known as craniotabes which, when found, is very characteristic; it consists of portions of thinned bone, causing a depression readily perceptible to the finger and confined to the occipital bone. Dr. Vogel considers this to be one of the very earliest signs of rickets.

Next comes a series of symptoms all more or less characteristic of the disease, and the first of them is profuse perspiration, especially of the head; the second is the desire to be cool at night, which leads to the kicking off of the clothes; the third is general tenderness of the whole surface of the body, so that the child dreads to be touched; and the last is an increase in the normal quantity of urine, which also is frequently loaded with phosphates. The child now looks old and careworn, it huddles up in its chair, it evidently desires to be let alone, the eyes have an unnatural brilliance, and the head enlarges. Simultaneously also the long bones are found to be curved and their ends to enlarge; this is well observed of the wrists, ankles, or ends of the ribs. M. Guérin points out that the deformities of the bones begin from below upwards; this is, however, not invariably the fact. The spine suffers similarly; if the child cannot walk, there is posterior curvature affecting the dorsal and lumbar vertebræ. If the child can walk the posterior curvature is dorsal, only that there is anterior curvature combined with it of the lumbar region. Moreover, these curvatures of the spine become associated with the flattening of the ribs laterally, which gives rise to "pigeon-breast." The teeth of rickety children are always backward, and when formed soon decay or drop out, so that "if a child pass the ninth month without teeth the cause should be looked for." Meanwhile the general symptoms deepen in severity; the child is more fretful if touched, and more dull, morose, or languid when left alone; its abdomen enlarges, its head enlarges, especially from before backwards, and it is also flattened on the vertex; and the fontanelle is depressed partly from debility, partly from the thickening of the cranial bones forming its edges. There is great loss of muscular power, the perspirations increase, the stools become very fetid, the appetite depraved and capricious, and if improvement do not take place from treatment the child sinks from exhaustion or from some thoracic or abdominal complication, such as bronchitis, or pleuritic effusion, or albuminoid infiltration of the spleen or lymphatic glands. This last condition is commonly associated with anasarca, and the child looks semitransparent, of a pale waxy appearance, not easily forgotten when once seen. The enlarged spleen can generally be felt, and the liver is not unfrequently enlarged also, though less commonly so than the spleen. Rickets may often terminate in chronic hydrocephalus, convulsions, diarrhœa, and laryngismus stridulus.

Should improved health take place, much of the superfluous bone is absorbed, the wrists and ankles get smaller, and, indeed, rickety children, if the disease be arrested, may grow up very healthy in every respect. The long bones, however, do not straighten, and the superfluous osseous tissue deposited along the cavity and at the ends of the bones is hardened, as some have said, by ordinary ossification, as others, and with more probability, affirm by a calcifying process, similar to that which occurs in enchondromata.

The muscular fibre in rickets is pale, almost transparent, and structureless, without any fatty degeneration. It is remarkable that in tuberculosis it may be scarcely half the normal size, and yet the child be able to walk and run; in rickets, with larger muscles, he cannot do either.

Sir W. Jenner has pointed out that the white patches often found on the left ventricle, a little above the apex, in children who have died of rickets, are produced by attrition, this being the very spot where the fifth rib bends inwards.

There is also usually found collapse of lung tissue, and that form of emphysema, called insufflation, due to overdistension, with air of the vesicular tissue of the lung. This condition invariably occupies the whole length of the anterior border of the lungs, extending about three quarters of an inch from the free margins. Between this emphysematous tissue and the healthy lung beyond lies a groove of collapsed tissue, which corresponds to those projections inwards of the ribs where they unite with the cartilage. The lymphatic glands, spleen, liver, kidney, heart and thymus, may, one or all, be affected with albuminoid infiltration. On cutting such an organ the surface is pale and transparent, compact, smooth, and tolerably moist. No reaction can be got with iodine.

The abdominal organs most frequently affected in rickets are the liver, the spleen, and the absorbent glands. The liver becomes dense, elastic, and pale, the cause of which, as Dr. Dickenson has pointed out, is increase in the portal fibrous tissue. The spleen is often enormously enlarged, hard and dense, purple in color, or mottled with buff color, on which the white Malpighian corpuscles show. The absorbents enlarge from hypertrophy of their cellular and corpuscular contents. The kidneys become large and pale, owing to an increase in the epithelium of their convoluted tubes. These conditions belong especially to the first four years of life, the osseous usually rather precede the visceral changes, but the visceral *may* precede the osseous, and may even be extreme, the osseous changes remaining but slight. And this is a clinical fact the recognition of which is of extreme importance lest our diagnosis be misled. Like the osseous changes, the visceral changes, unless extreme are recoverable and tend to recovery and do not impair the functions of the affected organs. For example, anæmia may result when the spleen is very large, but it is not usual to find either ascites or jaundice from any change in the liver, nor yet albumen in the urine from infiltration of the kidney. These visceral changes, have, however, one special clinical warning—when they have occurred the child is especially liable to the complications which accompany rickets.

Complications.—The diseases to which rickets predisposes and the occurrence of which in a rickety child will materially modify their ordinary treatment, the rickets, in fact, requiring ever to be borne in mind, are—

Coryza, bronchitis, pneumonia.
Diarrhœa.
Laryngismus stridulus.
Pertussis.
Convulsions.
Chronic hydrocephalus (rare).

Dr. Eustace Smith says, with regard to the influence of tubercle, rickety children may become tubercular, but a child in whom the tubercular diathesis is marked never, he thinks, becomes rickety. Rickets is rare also in the syphilitic diathesis. In this general freedom of intermixture of its type rickets comports itself like a true diathetic disease.

Treatment.—Remembering that improper feeding, bad ventilation, and "neglect" generally, are the great factors of rickets, so the treatment will comprise the reversing of these conditions; fresh air, proper food, and attention to cleanliness will alone do wonders. Children improperly fed on artificial foods should have a wet nurse, if possible, or be strictly dieted in quantity and quality of food. I believe that for rickety children it is good to add a little *salt* to their milk and water, or better milk and lime-water than sugar in any form. I have often found this to agree admirably when the ordinary milk and water and sugar has been constantly rejected by the stomach with every sign of acidity and dyspepsia.

The addition of a teaspoonful of cream to the half pint of milk and water is often beneficial.

Malt or barley food has been specially recommended for rickets. Four tablespoonfuls of ground malt should be boiled for ten minutes in a pint of water, the liquid poured off, and a pint of new milk added. The sediment from the husk if finely ground need not be removed, as it is very nutritious and rich in bone-forming material.

Ass's milk is a nutriment of great value in these cases, as also is goat's milk. Older children will require beef tea, bread, farinaceous and milk puddings, eggs, &c. In all cases fresh air and absolute cleanliness are to be insisted on. Tepid salt baths, followed by friction over the whole body with a coarse towel, are most useful; if the weather permit, the baths may gradually be made cold. Mercury, bleeding, blisters, and antimony, *are never to be thought of* for rickety children. Such children bear *mercury in particular* very badly. Of medicines I am sure there are none more valuable than cod-liver oil, combined with an equal portion of lime-water, the dose of oil not being too large for the stomach thoroughly to digest, that the bowels be not needlessly irritated, and the compound syrup of the phosphate of iron in drachm doses thrice daily. If preferred, phosphate of lime in small doses may replace the lime water. The syrup of the iodide and the old reduced iron are the next best forms in which to give steel. Alkaline medicines with a bitter infusion are occasionally needed to correct dyspepsia, especially "white stools" and acidity. The mineral acids are useful to check the extreme perspirations which sometimes occur. Tannin is recommended for this by Dr. Alison in doses of half a grain to a grain two or three times a day in a little dilute citric acid.

Catarrh and diarrhœa are the only two complications that need be noted here, though intercurrent disorders of whatever character occurring in rickety children will require restorative and by no means lowering measures. Catarrh must be at once attended to, lest it pass on to bronchitis or pneumonia. A linseed-meal poultice over the chest, a mixture of citrate and chlorate of potash, or citrate of potash and acetate of ammonia, is to be given, and if there be mucous râles or the tubes are evidently loaded with mucus an emetic to clear them. Powdered ipecacuanha or alum in solution is best for this purpose, and nourishing diet is not to be withheld on any account.

Diarrhœa, if merely painless without tenesmus, and the motions not specially offensive or discolored, will require astringents; if there be pain and griping, and the motions are offensive or discolored, especially if they be green or ochrey, a gentle purge, as rhubarb and soda, with some carminative, or a dose of castor oil in a little weak coffee or brandy and water, will be useful, to be followed by a bismuth mixture to which a little soda or potash may be added besides syrup and mucilage. If there be much pain and tenesmus a drop or two of laudanum may be added, and cinnamon-water will then replace the mucilage with advantage. These complications are so common and so important that the necessity for watching for them and meeting their earliest appearance by some such treatment as the above is imperative.

4. Syphilis.

The congenital variety is that derived from the blood of one or both parents. The acquired variety in infancy is that acquired by accidental contact with a chancre, as in a wet nurse or other person brought into close connection with the child, and extremely rarely by vaccination. Such a chancre resembles an ordinary chancre, and requires no further notice beyond bearing the possibility of its occurrence carefully in mind.

It will assist to ascertain if the syphilis be hereditary or acquired, if it occurred after or before the third month. Of 249 cases, 217 showed symptoms before the third month (Laucereaux).

The most striking symptom of congenital syphilis is "the snuffles," a snuffling noise which the infant makes in consequence of a subacute inflammation of the mucous membrane of the nose. If in addition to this the child is found to have condylomata about the anus, the diagnosis may be considered established. In addition, however, such a child is thin, poorly nourished; its muscles are not firm, but flabby, and its skin is brownish, cracked, thick, rough, and unwholesome looking. The fontanelle is usually open, and ossification proceeds but slowly. Nevertheless the teeth are forward rather than backward in their appearance. The posterior cervical glands are sometimes enlarged. The second set of incisor teeth often present a very characteristic appearance; the two centre teeth are short, narrow, and thin, the edges get worn away, and eventually they get broken, leaving a notch, and they are also striped or ribbed horizontally.

The child often looks prematurely aged; its hair may have fallen off, even the eyelashes and from the eyebrows; the corners of the lips and nose are often ulcerated, as also the margin of the anus, and it may have a symmetrical copper-colored eruption about the buttocks, &c., but especially characteristic when in the palms of the hands and soles of the feet. Progressive and general emaciation is a very unfavorable sign, the skin becoming of a dirty white, and hanging from the muscles, which are soft. In taking up a fold of the skin it is found to be inelastic, cold, and harsh (Nayler). Notwithstanding the emaciation the child seems always hungry, and often craves for food till the last.

The common syphilides are erythema, which occurs especially on the buttocks and perinæum in dark violet-colored blotches; lichen, psoriasis, impetigo, ecthyma, eczema, and pemphigus. The latter, when syphilitic in origin, especially affects the hands and feet. As a rule syphilitic eruptions are symmetrical, they are copper-colored, they are frequently arranged in a circular form, they are usually not accompanied with pain or itching. When ulcers occur their surface is ashy gray, covered with thick unhealthy-looking discharge, and the edges are red and sharply cut. Different kinds of eruptions, *e. g.*, papular, pustular, and tubercular, may coexist. In the squamous varieties, cracks and fissures are common.

The voice is hoarse and squeaking; the nails are small and badly formed, and liable to whitlows. The child suffers often from restlessness and sleeplessness, and has commonly a discharge from the ear. The liver is generally enlarged and hard, and has often undergone albuminoid degeneration.

The enlargement of the liver seldom causes jaundice, but ascites and œdema of the lower limbs are common. The enlargement is manifested by pains in the belly, vomiting, and frequently diarrhœa—more rarely constipation. The abdomen is tympanitic, and tender to the touch. Dr. Gee also cites enlargement of the spleen as occurring in about one half the cases. He goes so far as to say that an enlargement of the spleen is sometimes the only sign of an active syphilitic cachexia.

Iritis is rare, so also is excavation and ulceration of the tonsils, so also is disease of the bones, testes and brain (Berkeley Hill).

The disease is usually manifested in from fourteen days to six weeks from birth, or, more rarely, the child may be born with it, or, more rarely still, the symptoms do not appear before the seventh or fourteenth year. Mr. Nayler states that of seventy-three cases only four could be enumerated as occurring within twelve days from birth. It is of the utmost importance that the syphilitic diathesis should be early recognized, as all the after management of the child in health and disease will be modified by the fact of its existence.

Mr. Naylor asserts that the development of secondary in contradistinction to tertiary syphilis, contracted a few months before marriage, and from which the parent has imperfectly recovered, is one of the most frequent causes of syphilis affecting the child. He mentions a case where the father, suffering from syphilitic eczema, infected his child, though an interval of thirty years had elapsed since the accession of the primary sore. Tertiary syphilis seldom leads to congenital syphilis, or at any rate, to its manifestation in early life.

Treatment.—Mercury in some form or other is the great sheet-anchor in this disease. Practitioners vary in the method of its administration. Some prefer the use of gray powder, in doses of from half a grain or two grains twice or thrice daily, with a few grains of compound cinnamon powder to prevent the mercury running off by the bowels. The cinnamon powder may be replaced by P. Ipecac. c. Opio when necessary.

Many recommend the use of Ung. Hydrarg. (ℨ ss.) rubbed into the thighs or arms, or simply laid on by means of a piece of flannel. When it is desirable to avoid suspicion I do not think these plans are good, and in out-patient hospital practice the mothers are often too ignorant or too careless properly to apply them. In the case of a sucking child it is recommended to give the mercury to the mother, and this is often attended with undoubted benefit; but in severe cases it will usually be found necessary to give a little separately to the child. The mercury, in whatever form given, will require to be persevered with for about six to twelve weeks. It is often useful to change the method during this time, and then the combination of corrosive sublimate with bark is especially useful. To an infant under four weeks five drops of the Liq. Hydrarg. Perchlor. may be given twice a day in a little Infusion of Bark, to which a few drops of glycerine, and, if necessary, half a drop of Liq. Opii Sedativ. may be added.

Vomiting and diarrhœa are indications for temporarily stopping the mercurial course, or at least for changing its form. Sometimes a little chalk or carbonate of potash added to the gray powder enables it to be well borne. Sometimes the external use only can be tolerated for a time. Mr. Dann advocates the use of chlorate of potash where mercury disagrees —about four grains is the appropriate dose—twice or three times a day in well-sweetened water.

Black wash is the best local application; the anal condylomata should be dusted over with calomel, and kept scrupulously clean. Occasionally they may require a touch with nitrate of silver. The yellow oxide of mercury, four grains to one ounce of benzoated lard, is a useful non-irritating application. Caustic is also useful in ulcerations of the mouth and tongue. Iodide of potassium, bark, and sarsaparilla are of value after the mercurial course in improving the general health. The Syr. Ferri Iodidi is a favorite medicine for this purpose. Baths of corrosive sublimate are sometimes useful in the skin complications. If difficulty occur in dieting, goat's or ass's milk may be resorted to. A wet-nurse cannot be medically sanctioned; there is risk of contagion from which the mother is, however, free, as has been proved by many observers. I have known one or two cruel instances in which a wet-nurse has been infected. Such cases should be punishable by law.

5. Acute Rheumatism

is an affection not very common, especially in young children, but inasmuch as slight attacks of it, even such slight manifestations of the diathesis as torticollis and erythema nodosum, are sometimes complicated with pericarditis, its consideration is important. Rheumatism is an occasional complication of scarlatina, and it is then more especially liable to be attended with cardiac mischief. The disease, when it occurs idiopathically, sets in usually with rigors and feverishness, followed in a day or two by the swelling of the joints destined to be affected. The fever commonly runs high, not so high, however, as in adults; a temperature above 104° F. indicates a grave attack; the tongue is loaded with thick white fur, and the body is bathed in a peculiar acid perspiration of an odor which when once noticed is always remembered. The urine is scanty and high colored, and loaded with lithates. The large joints are generally the earliest affected, and the disease often wanders from ankles and knees to elbows and wrists, the swelling readily subsiding in one set as another is attacked. A joint afflicted with rheumatism is reddish, swollen, very tender, and very hot; and a joint presenting all these symptoms urgently one day may on the next or day after be almost free from symptoms. The disease lasts from ten to thirty days (seldom more than fifteen), and is usually subject to exacerbations at night. Inflammation of the pericardium, which, especially in children, occurs in a very large percentage of cases, is manifested by tightness and slight pain in the chest, often so slight as to be overlooked unless narrowly watched for, and besides, in children especially, delirium. Friction sound becomes audible at the base usually, but often over the whole area of the heart's dulness. As effusion takes place, this natural area of dulness is increased, and the sounds of the heart are muffled by the intervening fluid when the disease affects the lining membrane of the heart (endocarditis).

The murmurs are caused at the base or apex, according as they are aortic or mitral in origin; they are usually systolic; occasionally, however, a basic diastolic murmur is heard. When the heart substance is attacked (carditis), great irregularity of action takes place, with syncope and often fatal collapse (see Pericarditis and Endocarditis).

Dr. Vogel records a case of well-marked acute rheumatism and endocarditis in a child one year and nine months old. It is much more commonly seen in children who have passed the sixth year. A child suffering from heart disease following rheumatism may live for many years. I have been frequently surprised to observe little ones running about whose span of life I had deemed brief indeed. Dilatation and hypertrophy with a mitral murmur is a common condition to find; heaving impulse, with increased area of dulness on percussion, are present. The child suffers from frequent and violent attacks of palpitation if allowed to play about; but later on, such attacks occur without exertion, and are attended with much gasping for breath, and the child has to be propped up on pillows at night. Dropsy is not a marked symptom even in the advanced stage of such cases, but it may occur from time to time, especially if a little bronchitis or other cause add to the heart's difficulties in driving forward the column of blood. A condition varying in detail from this may go on for years, but despite all the relief we can afford, such children rarely attain maturity.

Treatment.—The affected joints should be wrapped in flannel or cotton-wool, and hot poppy fomentations applied, or the joints may be covered with extract of belladonna. Lotions of carbonate of soda have their advocates. Blisters in children are decidedly undesirable. I am accustomed often to rely on the so-called alkaline treatment, remembering that the materies morbi is by many attributed to excess of lactic acid in the blood (e. g. Dr. B. W. Foster has recently produced all the symptoms of acute rheumatism synthetically by the administration of lactic acid), and that the urine is (if no alkalies be given) unquestionably very abnormally acid. The bicarbonate of potash with the citrate of potash in fair doses, ten to fifteen grains every four hours, seems decidedly to relieve pain and to render the disposition to cardiac mischief less likely. Lemon-juice, iodide of potassium, guaiacum, colchicum, mercurials, and vapor baths, have all their supporters. Iodide of potassium is of value in some cases, especially when the more acute symptoms are passing away; in combination with bicarbonate of potash and infusion of serpentaria, there is no more valuable remedy for rheumatic pains, which are so commonly the sequelæ of the acute disease. Colchicum can rarely be needed in children. Cimicifugin in small doses, gr. ½—gr. ij., according to the age of the child, is a most valuable remedy when there is a tendency to chorea, which is by no means an uncommon accompaniment of rheumatism.

Salicin is attracting much attention, and is most highly spoken of in acute rheumatism. The more acute the case the more valuable the remedy, and its value is shown rapidly, within forty-eight hours of its administration. It produces marked results, such as relief of pain and fall of temperature. A moderately large dose, say 10 grains for a child ten years old every three hours, is recommended. It is averred that cardiac complications are commonly prevented, and that until the temperature becomes normal, a beneficial action in a heart disposed to inflammatory action is exerted. When the normal temperature is reached the value of the drug ceases. Pure salicylic acid is also similarly employed, but is stated to be more irritant in its action, causing sickness and diarrhœa occasionally. From the undoubted testimony in their favor these remedies are worthy of further trial. Relief for the elevated temperature may also be obtained by the "cooled bath" (see Typhoid Fever and Baths). Dr. Binz and other writers speak highly of the value of cold baths, e. g. 68° to 70° F. These are recommended to be given for a short period and often repeated. The value of aconite as in other acute inflammations must not be lost sight of. More will be said upon this matter hereafter. It is certain that many cases do well upon drop doses of aconite given at the earliest indication of rheumatic fever. Copious perspiration with relief of pain and fall of temperature are observable. It needs the intuition and confidence of experience to adopt one or other of several excellent lines of treatment, but the "cooled baths" may be used as an adjunct to either the alkaline, aconite, or salicin plan.

The bowels will require to be carefully regulated, and a little P. Ipecac. Co. at night is useful either as a sedative or as a diaphoretic. Great relief is often obtained by sponging with warm water at bedtime, carefully wrapping the affected joints afterwards in cotton-wool. When cardiac mischief has unhappily occurred, a few leeches over the heart give relief sometimes, and are preferable to blisters. If a blister be thought necessary, notably from great effusion, it should be effected by blistering fluid, and not by the Emp. Lyttæ, which last in delicate children often leaves the most intractable sores. Small doses of calomel and opium are of value. The child must be kept absolutely quiet, and if the heart's action be much excited a belladonna plaster externally is useful. There is no occasion, and indeed it is better practice not, to discontinue the alkaline treatment during cardiac inflammation. But the remedy of all others most serviceable in chronic cardiac affections is digitalis. It is hard to say what digitalis will not accomplish in heart disease. It will improve the pulse, it will remove dropsy, it will prolong life for years. When much hypertrophy exists aconite may be combined with it. If the pulse be irregular (a rare thing in children) belladonna may replace it for a time.

In these three drugs the physician has to his hands agents of the utmost value in managing the many variable features of chronic affections of the heart.

The diet throughout must be light, yet nutritious. Soda-water and milk in equal parts is very grateful. Beef-tea and broths may also be given. Stimulants are needed, chiefly for heart complication, when the system evidently flags. Thirst may be allayed by cooling drinks, as lemonade, toast and water, currant-water, weak tea, imperial, &c.

CHAPTER III.

SKIN DISEASES.

1. EXANTHEMATA.

Roseola (Rose Rash).—A mild, non-contagious, subpyrexial disease, characterized by small rose-colored spots or transient patches of redness, which endure from twenty-four hours to a few days or a week. The especial form of it now chiefly to be considered is *R. infantilis*, sometimes called false measles. This is especially common in hot weather, attacks the most prominent parts of the face and extremities, or may be limited to one limb. The eruption itches slightly. It lacks the crescentic form and the constitutional symptoms of measles. It is deep rose, and not scarlet like the rash of scarlatina, neither can it be resolved into the innumerable red points which go to make up the scarlatinal eruption, nor are the patches so uniform nor so diffused as in that disease. It is often preceded by sore throat. Roseola sometimes precedes variola, scarlatina and rubeola; it then shows itself chiefly at the joint flexures. I have more than once seen a child affected with sore throat, and undoubted roseola patches disappearing in twenty-four hours, whilst one of its brothers or sisters has been suffering in another room of the same house from scarlet fever or measles, and it has occasionally happened that the child attacked with this transient roseola has escaped the severer affection altogether. The treatment of these cases is very simple; a moderated diet, mild laxative salines, simple drinks, and a warm bath or two, generally suffice for the cure. If the disease appear due to swollen gums and the irritation of teething, the gum lancet should be freely used, but not otherwise, as has elsewhere been frequently mentioned.

Erythema (Intertrigo, Red Gum, Tooth Rash) is known by large, slightly raised, red patches of irregular form, disappearing on pressure. The disease is non-contagious. It is common in the rheumatic diathesis; the eruption itches and burns a little. It is known from roseola chiefly by the rosy tint of the latter. It may be of local origin, as when two folds of skin rub against each other, as in the necks and loins of infants; the surface is then moist. Such intertrigo is best treated by washing the child with warm water, using no soap, and after carefully drying by dabbing with a soft cloth, dredge oxide of zinc, fine starch, or lycopodium powder over the inflamed surfaces. A mild aperient will complete the cure. If the disease be constitutional it may occur as—

E. nodosum, especially common in young girls. This eruption consists of red, elevated, oval patches, from one to two inches in length and about three-quarters of an inch wide. Each patch lasts from four to ten days, and fresh patches continue to appear; the whole disease may last three or four weeks. It occurs generally on the fore part of the leg. The color, at first red, passes into a bluish tint. It is usually preceded by slight fever and other constitutional disturbances. These patches, even when several coalesce, do not suppurate, though an obscure sense of fluctuation may thereby be imparted to the finger. The diagnosis from erysipelas is made, not only by the greater amount of constitutional disturbance, but also in erysipelas the margin of the patch is as much raised as its centre, the hardness is superficial and brawny, and its redness ends in a well-defined line; whereas in erythema these conditions are reversed. The disease is called E. papulatum and tuberculatum, according to the size of the patches. Chorea and rheumatism are sometimes associated with erythema nodosum.

Treatment.—Mild aperients, warm baths, light diet, and attention to the teeth and digestive organs. Bark and quinine are useful in E. nodosum; steel is indicated if, as often, it is connected with chlorosis and amenorrhœa.

Urticaria (Nettle Rash).—In addition to red patches fading on pressure, this rash presents also wheals such as might be produced from the sting of a nettle or the stroke of a whip; there is intolerable itching and irritation, which is aggravated usually by the heat of the bedclothes, by the heat of a fire, or by the use of wine and stimulating condiments. It is non-contagious; the disease may be acute or chronic. Vomiting and diarrhœa seem to be the natural conditions of its cure.

It is caused by indigestible diet, acting, of course, on a constitution predisposed to take the disease; such diet is often shell-fish, mushrooms, cucumbers, certain fruits, pork, and also some medicines, as turpentine and copaiba. I saw some years ago a most remarkable case, in which copaiba was the cause, and in which the eruption was of the most extensive and formidable description, ushered in with smart fever, extending over the whole body and resembling scarlet fever in its appearance in many respects. This case yielded readily to a calomel purge, followed by salines and diaphoretics, and in thirty-six hours the patient was well. The patient was a woman of about thirty years of age.

Urticaria is more often a chronic disease, and it then forms several varieties, as conferta, perstans, evanida, and tuberosa. These varieties are rare in children, in whom urticaria is more frequently acute or dependent on teething or improper food.

Treatment.—If acute, emetics and purgatives, to imitate the natural cure; the gums should be lanced if at fault. If chronic, the diet must be regulated, stimulants avoided; the local irritation may be relieved by sponging the surface with vinegar and water, or a lead lotion, or a lotion of prussic acid and glycerine, with almond emulsion, or by the use of alkaline baths. The most reliable drugs are arsenic (the Liquor Arsenicalis in small doses immediately after meals), quinine, and occasionally the alkalies to correct acidity. For chronic cases Dr. Tilbury Fox recommends sulphuret of potassium baths ℥j. or ℥ij. to 30 gallons of water. He employs this bath two or three times a week, and also uses an oxide of zinc lotion, to which is sometimes added a little Liquor Carbonis Detergens (℥ij. or so to ℥vj.).

Erysipelas.—The infantile variety, or that occurring in children under six months, starts usually from the navel; occasionally it follows on intertrigo, and very rarely on vaccination. Infantile erysipelas, like ordinary erysipelas, presents redness, œdematous swelling, heat and pain. The redness may deepen to lividity, and vesication or even gangrene occur. It is always a very dangerous disease, and may prove fatal in seven days or less; more often, however, the duration is two or three weeks. The most marked diagnostic character of infantile erysipelas is the disposition to wander from surface to surface by insidious and creeping migration, often until the entire body has been attacked.

The disease is fortunately rare; it is especially liable to occur during an epidemic of puerperal fever or under especially depraved hygienic conditions of the lying-in room. The navel in such cases never heals kindly, and umbilical phlebitis is most frequently the proximate cause of the erysipelas.

Treatment is sufficiently unsatisfactory. Bouchut says *all* new-born infants attacked die, and that even up to a fortnight one rarely recovers. Tincture of the chloride of iron, two drops in sweetened water every two hours, bathing with warm mucilaginous fluids, and dredging in the intervals with unscented starch or flour, and pencilling round the disease with nitrate of silver on the sound skin, are means ordinarily recommended. Brandy ten drops, with a little aromatic spirit of ammonia well sweetened, may be given when there is much prostration. The infant should, of course, have the breast, or if weaned, milk and beef-tea will be the best nutriments. A lotion of sulphate of iron (℥ ss. to ℥ viij.) is highly commended by Velpeau. A lotion of carbolic acid with glycerine is of use where fœtor is present from sloughing and the exudation of thin sanious pus. Extreme support and stimulation afford the only hope in cases where sphacelation occurs, and the first sign of improvement is the appearance of "laudable" in place of the thin ichorous pus. Free incisions should be made when needed to give issue to collections of matter, and a charcoal or yeast poultice is then a nice application. Quinine, a grain or two, two or three times a day, may be tried. If constipation be present the bowels should be relieved, but there is more often diarrhœa, which which may even require Dover's Powder or opium to quiet. It is almost needless to add that no treatment whatever has the slightest chance of success if the hygienic conditions, especially in the matter of fresh air, be allowed to remain faulty. Dr. Ringer extolls aconite in the erysipelatous redness after vaccination, and also in erysipelas itself.

2. VESICULÆ.

Sudamina are colorless and transparent vesicles, and resemble drops of perspiration; they appear in the course of acute rheumatism, typhoid fever, &c.; their contents are acid; they are of no clinical signification, and require no treatment.

Miliaria.—Miliary vesicles are pointed and have a red blush around their bases, they become opalescent and even purulent in appearance; in all these points they differ from sudamina. They occur also in the course of acute rheumatism. Children who perspire freely, and especially if seldom washed, often become covered with them. The treatment will be thorough and frequent ablution and a mild aperient.

Eczema is inflammation of the sweat-follicle according to some writers, and catarrhal inflammation of the skin according to others; it consists of vesicles surrounded by a red zone; the zones coalesce and the vesicles burst; a fluid exudes, which is alkaline. When the vesicles have burst, scabs or scales cover the surface of the patches. It is non-contagious. Parts affected with eczema smart and burn rather than itch. A common form in children is eczema impetiginodes; in this variety the inflammation is exceedingly acute and is accompanied by swelling; the contents of the vesicles become purulent, and yellowish scabs form. In children the scalp and the ears are, perhaps, the most frequent sites of eczema. In strumous children it occurs often at the flexures of the elbow- and knee-joints. Eczema generally chooses situations where the skin perspires most and is soft and thin, as the inside of the limbs, the flexures of the joints, &c. Mr. Balmanno Squire has pointed out that when the eczematous patches are confined to the occipital portion of the scalp they are usually caused by pediculi, whereas the constitutional variety affects the anterior portion of the scalp, a statement quite borne out by my own observations.

The syphilitic variety is of sufficient frequency and importance to call for a few words. In it there is not so much discharge or crustine, but more swelling, generally some induration, and sometimes darkish looking scabs. These cases will not benefit without mercurial treatment, notably the perchloride and bark.

Treatment.—If acute and inflammatory, a free purge of calomel and jalap, followed by saline aperients, will be useful at the outset. If chronic, gentle aperients may be resorted to; the diet must be carefully regulated, and the benzoated oxide of zinc ointment may be locally applied; the gums should be lanced if they require it. If the case be of strumous origin, cod-liver oil and steel will be necessary, and the best application is then a lotion of nitrate of silver (gr. xx.— ℥ j.) applied on lint twice a day. In eczema capitis it is always better to remove the scabs by poultices of linseed meal before using the zinc ointment. If the case be very obstinate, and there be no heat and swelling, linseed oil having been applied at bedtime to the scalp (freed first from hair and scabs), the part may be covered in the morning with liquid pitch; this is a strong measure and requires care, but it is usually very efficient. It is necessary that the skin be kept very clean; for this purpose juniper tar soap will be found useful; common soaps, especially scented soaps, are better avoided. It is sometimes useful to employ an alkaline lotion—Sodæ Bicarb. 3 j., Aq. Oj.; also in chronic cases small doses of Liq. Arsenicalis three times a day; if this, however, should cause diarrhœa, it must be left off. It is generally most serviceable to combine its use with iodide of iron and cod-liver oil. In obstinate cases calomel ointment or the Ung. Hyd. Nit. Mitius, or a lotion of nitrate of silver, may be tried. In some extremely chronic and obstinate cases I have been much pleased with the effects of the Oleum Rusci (or Oleum Betuli Albi). I generally dilute it with an equal part of pure glycerine, and have the application used every night. The effect in some cases has been almost magical; in others good has resulted, but in a slower and more ordinary way. I have never employed the oil of white birch without a previous course of either alkaline or arsenical treatment (with an occasional aperient) for a fortnight or so.

In the case of a young lady about 16, a sufferer for years, whose case had resisted all previous efforts, *one week* of Oleum Rusci established a cure, which remained for six months afterwards, when I last heard of her, nor did any unfavorable symptoms appear. The powerful smell is, of course, a disadvantage, but it is not worse than other "tarry" preparations. Of Gurjun oil I am sorry to say I cannot speak favorably; I have tried it pure and diluted with an equal part of lime-water, as recommended by Erasmus Wilson, but the results have not been satisfactory in my hands. In one case very smart inflammation was set up. I have heard of cases being benefited by its use, but have had no opportunity of verifying the statements. When the disease is caused by pediculi, or by scabies, the cure of these conditions is sufficiently indicated.

Herpes (Tetter) is a non-contagious disease, consisting of clusters of vesicles of spheroidal shape and of some size, upon inflamed patches of irregular form. The vesicular contents, at first watery and neutral in reaction, soon become yellowish-white or purulent, and escape and form scabs. The disease may last a week or ten days. The simplest variety is H. labialis, which appears usually on the upper lip during a common cold; it often occurs also in mild cases of lobar pneumonia.

Herpes zoster, or Shingles, is common in children. This variety is ushered in with poorliness and subpyrexial conditions, and occupies half the body, usually the right side, in the form of a band; the thorax and lower part of the back and groin are the commonest sites. There is generally some pain in the part before the rash appears. The treatment is a mild saline aperient, plain diet, and warm bath. If any local application be used, warm mucilaginous fluids are the best, but they are seldom needed.

H. circinatus occurs in two forms: that with large vesicles runs the usual course of herpes, and requires the same treatment as H. zoster.

That with the small vesicles is arranged in rings somewhat smaller than a threepenny piece, with a centre of sound skin and a red border. The disease spreads at the circumference and heals in the centre. It is furfuraceous and without constitutional derangement. It has a tendency to become chronic, and often occurs in the strumous child. This form of the disease is contagious. Local applications are the appropriate remedies; as strong solutions of sulphate of iron or gallic acid. An application of acetic acid, or of a solution of nit. silver (℥ j.— ℥ j.) will suffice when milder measures fail. Hyposulphite of soda lotion (℥ iv. ad Aquæ ℥ vj.) or sulphurous acid lotion (℥ j. ad Aquæ ℥ iv.), or ammonio-chloride of mercury ointment (gr. v. ad ℥ j. adipis) are all very serviceable when the disease is in schools, and it is desirable to destroy all chance of contagion as quickly as possible. Touching the patches with Liquor Epispasticus is another rapid method of cure. Tonics and cod-liver oil and enforcement of good hygienic principles are necessary, as in so many other affections of the skin, parasitic and otherwise. Some writers assert that herpes circinatus (or, as it is sometimes called, tinea circinata or ringworm of the surface) and tinea tonsurans are the same disease from their being frequently associated. The balance of evidence, however, seems to favor the opinion that herpes circinatus forms a nidus eminently favorable to the propagation of the spores of the *Trichophyton Tonsurans*. (See Tinea Tonsurans.)

3. BULLÆ.

Pemphigus, Pompholyx.—This eruption is usually preceded for twenty-four hours or more by a feeling of lassitude, sickness, and headache; feverishness, and even delirium, are not rare; there appear a number of clear blisters varying in size from a sixpence to a half-crown. The blisters rest on an inflamed patch, which may or may not form an areola around them; they appear on the face, neck, trunk, and extremities. In a few days they attain their full size and either fade away or burst. The bladders shrivel and leave some brownish-colored scabs. The duration is from one to three weeks. When seen on the soles and palms of young children pemphigus may be regarded as certainly syphilitic in origin.

The disease often attacks young and poorly nourished infants. The variety called infantilis may, however, also be produced by dentition, bad feeding, or any gastro-intestinal irritation. The disease shows a tendency to become chronic in enfeebled children.

Pompholyx, which is rare during childhood, is merely a variety of pemphigus unattended with fever. Pemphigus is always a serious disease, and, when chronic, it is very obstinate.

Treatment.—The vesicles should be punctured as they appear, and if there be any difficulty in the separation of the scabs they should be poulticed, and the ulcerated surfaces beneath treated with some stimulating application; nitrate of silver or weak nitric acid lotions are perhaps the best. If the disease occur in a weak cachectic child it will require bark and quinine, stimulants, and good food, for its cure. If, on the other hand, the type be inflammatory and the child be strong, which is rarely the case, moderate doses of saline aperients, low diet, and tepid drinks, will be appropriate. In the chronic form it is well to give iodide of potassium a trial, especially if the disease be suspected to be syphilitic. Arsenic (the Liquor Sodæ Arsenitis Mr. Naylor considers the best form for children) may also be used, and cod-liver oil is frequently beneficial. Alkaline and gelatine baths have been recommended by different writers.

Rupia.—The bullæ in this disease are round, flattened and isolated, about the size of a shilling; filled at first with serum, which soon changes to pus. These bullæ are surrounded by an inflamed areola; after a time the bullæ shrink and become covered with thick, brownish scabs. One variety—rupia prominens—resembles a limpet shell, the scabs being thicker and formed by several layers of hardened secretion consequent on the extension of the ulcerated surface beneath the scab.

Rupia usually occurs in the lower extremities; rupia prominens is almost always syphilitic; rupia simplex, the least virulent form of this disease, and rupia escharotica, which is accompanied by much constitutional disturbance and extensive ulceration, often with sloughing, are both common in children; the former in children about six or seven, the latter in infants up to the period of the first dentition; it is then a very dangerous disease. Rupia escharotica generally appears on the genitals, the legs, or the scalp.

Treatment.—The bullæ should be punctured as soon as formed, the scabs, if obstinate, may be removed, and the ulcerated surfaces below dressed with solution of nitrate of silver, black or red wash, or some stimulating ointment. Alkaline and gelatine baths are recommended. The constitutional treatment should be tonic; if the disease be syphilitic, iodide of potassium; if otherwise, quinine, the mineral acids, or bark, with generous diet and wine. In R. escharotica it is sometimes necessary to have recourse to powerful caustics (the acid nitrate of mercury is the best), and the eschar should be coated with calamine in glycerine, or with collodion.

4. PUSTULÆ.

Impetigo (Running Tetter) is a suppurative inflammation of the hair follicles. It is contagious. This common eruption appears in the form of clusters of small pustules slightly raised above the skin, which burst in a few days, exude slightly acid purulent fluid, and harden into thick, yellowish scabs. There is more or less constitutional derangement, with great heat and itching of the parts affected. The lymphatic glands in the neighborhood always enlarge even in slight cases, and in strumous children will suppurate.

Impetigo Figurata occurs usually in patches of definite form on the cheek. This is a common variety in children about the period of dentition; children also often suffer from the disease on the scalp, which is called impetigo capitis. It is of importance to remember that no permanent disfigurement of the skin remains even after severe impetigo. The disease is often obstinate and shows a tendency to become chronic. The treatment will comprise a moderate and well-regulated diet and saline aperients; locally, the crusts should be removed by steaming them, or by means of a poultice; if on the head the hair must be cut short, and the operation of removing the scabs will be difficult and require patience. When accomplished the Ung. Zinci rubbed in twice a day is often sufficient to cure the disease—if it fail, however, and fresh crops of pustules continue to appear, the next most useful application is Ung. Hyd. Nitrat. Mitius. This rarely fails where the zinc has not succeeded. Should, however, a case be extremely obstinate the Ung. Sulph. Iodid., or especially the Ung. Hyd. Ammon. Chlorid., or the Ung. Hyd. Nit. Oxyd., will succeed; at the same time the little patient should be freely purged from time to time, and take regularly from half a grain to a grain of quinine twice daily; sometimes steel will seem more beneficial. Cod-liver oil is also valuable in the strumous and ill-nourished. Very chronic cases require the use of Liq. Arsenicalis. Impetigo capitis, like eczema, is frequently caused by pediculi.

Ecthyma.—This eruption consists of pustules formed in the centre of inflamed circumscribed patches; in a few days the pustules dry up and leave thick scabs; when the scab separates purple discoloration of the skin or in strumous cases an unhealthy ulcer is left. The disease is generally chronic; children are very prone to suffer from this disease; it is not contagious, but it is not unfrequently, especially in children, accompanied by scabies. It is sometimes of syphilitic origin.

Emollient applications, of which warm water is one of the best, and simple saline aperients, with a moderately generous diet, form the treatment of the acute disease. If chronic, more stimulants will be required, the ulcers may be touched with nitrate of silver, or a stimulating ointment employed. A useful application is one drachm of Tinct. Benzoin. Co., to one ounce of lard. Bark and steel and cod-liver oil will also be of service; when scabies is present it must be attacked at once.

5. PAPULÆ.

Lichen.—Lichen is an eruption of small, hard red pimples, which do not disappear on pressure, and which itch greatly. There is no discharge or weeping of any kind. The eruption usually terminates in desquamation. The commonest sites are the back of the hands, forearms, trunk, and face. It is generally a chronic disease and not contagious.

Lichen strophulus or Red Gum is a common affection of infants from birth to the first dentition. It generally lasts three or four weeks and has many varieties, according to the color of the skin and the arrangement of the pimples. It usually affects the face, neck, and hands. It requires merely a little rhubarb and magnesia, attention to the diet, and lancing the gums when necessary.

Lichen simplex.—The papillæ are the size of millet-seeds and bright red; it is often pyrexial and subacute.

Lichen urticatus has papules of a large size, preceded by wheals, which look like the sting of a nettle. It is a complaint common to early dentition, especially in hot weather.

In *Lichen agrius* the papules are confluent and on an inflamed base; the pain and smarting are severe, and considerable constitutional disorder is manifested. After a while the papules burst, sero-purulent fluid is exuded, and painful cracks are formed; this variety may last from ten days to several months; it is not common in children.

Treatment.—Careful attention to the diet and to the bowels will be required in all the varieties. Stimulants and excitement must be avoided, the child must be warmly clad and should be bathed every other day in a gelatine bath; acidulated drinks may be given. Of local applications, the best are those which relieve itching, as Liquor Plumbi Diacetatis, prussic acid, or

℞ Glycerini... ʒij.
    Hyd. Bichlor....................................... gr. vj.
    Chloroformi.. ℳxx.
    Aquæ... ʒvj.
Ft. Lotio.

Or, pyroligneous oil of juniper, and spirit of wine, of each one fluid ounce, to be added to six ounces of water, are useful lotions. Fowler's solution and iodide of potassium with serpentaria are useful in very chronic cases, for the disease is often syphilitic.

Prurigo.—In this disease the papulæ are the color of the skin, larger than those of lichen, and very chronic in their course; moreover their itching is well-nigh intolerable. In children the disease is seldom so severe as in the aged; the commonest site is the neck and shoulders. When the spots have been scratched they often present a small dark spot of blood on their summits.

The best *treatment* in children, after regulating the diet and the state of the evacuations, is the daily use of tepid alkaline baths. Sponging with vinegar and one of the lotions to relieve irritation, and the internal administration of nitro-muriatic acid in sarsaparilla; taraxacum in some cases, and Fowler's solution in others, act like specifics. Quinine and cod-liver oil are valuable in the strumous. Sulphur vapor baths are also useful in many instances.

C. Squamæ.

Psoriasis (Dry Tetter, with which is included Lepra vulgaris) is a chronic, non-contagious disease, characterized by slightly raised red patches covered with silvery scales, and without itching. The eruption has many varieties, according to the site, extent, and course. The red patches are due to inflammation of the skin, the scales to excessive formation of epithelium on the inflamed surface. Its commonest sites are where the skin perspires least and is coarsest; hence below the knee and on the elbow are characteristic positions. If it occur on the palms and soles it is usually syphilitic in origin. The disease is not very common in children.

Treatment.—Local applications are of little service, but warm baths and alkaline baths are useful. Liquor Potassæ, Pot. Iodid., and especially Liquor Arsenicalis, are valuable remedies. Tincture of cantharides has been recommended by some writers, especially Cazenave. Arsenic will be required to be given with the usual precautions, and the decoction of dulcamara will form a good vehicle for its administration. Some cases require mercury for their cure; the perchloride is the best preparation, and it should be given with bark. Carbolic acid, two grains to one ounce of water, and a little glycerine, is useful in allaying itching; the Liquor Carbonis Detergens (Wright) has a similar use, diluted pretty freely. Mr. Nayler recommends a touch with blistering fluid, or glacial acetic acid, immediately to be followed by a touch with a weak solution of soda, to diminish excessive irritation, should such be excited by the blistering fluid. This should only be resorted to when the disease proves chronic and obstinate.

Pityriasis (Dandriff) is chronic inflammation of the skin, attended with itching and abundant desquamation of small scales or scurf. The head is the part commonly affected. If the disease becomes chronic the hair gets thin. Pityriasis capitis is not uncommon in newly born infants; there is no constitutional disturbance. Pityriasis rubra and versicolor are varieties chiefly marked by their color, the former red, the latter yellowish-brown. The best application in Pityriasis capitis is a lotion containing borax and Tinct. Arnicæ; sometimes the citrine ointment is of service; the head should be kept very clean, and a soft hair-brush used. A tonic aperient will complete the cure.

7. TUBERCULA.

Acne is inflammation of the sebaceous follicle, with accumulation of its secretions.

This disease and its varieties will not detain us, as it does not occur in childhood in any form.

Molluscum is enlargement of the sebaceous follicles, and presents indolent tumors, varying in size from a threepenny piece to half-a-crown, sessile or pedunculated, of the natural color of the skin, and containing an atheromatous matter. At the same time there is neither constitutional disorder, nor pain, nor ulceration. The trunk and head are perhaps the commonest sites, but they may appear anywhere. The disease often occurs in childhood, and lasts for years or throughout life. It is contagious. The treatment is to slit up the tumors and touch them freely with caustic, if necessary, or if pedunculated they should be cut off, and the base from which they spring cauterized.

To this order also belong Elephantiasis, Keloid and Framboesia, of which disease Elephantiasis and Framboesia occur only in foreign climates, and none of them occur in children.

Lupus is also an affection very rare in children; there are two varieties:

1. *Exedens*, which is highly ulcerative, and destructive in character, and attacks especially the nose.

2. *Non-exedens* is without ulceration, but the tubercles leave deep pittings and marks of cicatrization, and also the skin is seamed by white scar-like ridges, very characteristic when once seen.

Donovan's solution, or Liq. Arsenicalis internally, and one of the powerful caustics, such as chloride of zinc, potassa fusa, or nitric acid locally, form the treatment of L. exedens. Of L. non-exedens the constitutional treatment will be the same, but the caustics need not be so powerful; acetum cantharidis, the iodide of mercury, and iodide of sulphur ointments answer best; the disease is often syphilitic.

8. XERODERMATA.

Ichthyosis, or Fish Skin, is often congenital; the skin is dry, harsh, and rough; in a further stage the whole body, or the palms and soles, face, eyelids, outer surface of limbs, &c., are covered with small, hard, thick, dry, brown scales, overlapping each other like the scales of a fish; there is no heat, pain, nor itching. Patients afflicted with this disease are usually cachectic, and have often a disagreeable smell. The disease is generally hereditary, if not congenital. The treatment will comprise alkaline baths, and subsequent inunction with glycerine, almond oil, or elder flower ointment, arsenic and cod-liver oil may be tried. The disease, however, is often incurable.

9. Parasitici.

The vegetable parasitic diseases are Tinea tonsurans, favosa, decalvans, and Chloasma. The animal parasitic disease is Scabies.

Tinea tonsurans (Porrigo scutulata, Ringworm). This disease occurs in patches of circular form, in size from sixpence to half-a-crown, on the scalp. It is chronic and contagious. The surface of the patch is covered with loose white scales, while the hairs look as if they had been cut off close to the patch. This is because the hairs are very brittle, and they also possess no elasticity. The vegetable parasite causing the disease is called *Trichophyton tonsurans*, the spores of which infiltrate the hair. Herpes circinatus is by some writers regarded as being identical with this disease. Sir W. Jenner maintains that the diseases are quite distinct, but that the secretions of the part of the skin affected with Herpes circinatus form a favorable nidus for the growth of *Trichophyton tonsurans*. Tinea tonsurans affects not only the scalp, but the neck, trunk, &c.

Dr. T. Fox, who has published several valuable papers upon Ringworm in schools, insists upon the necessity of disinfecting the air, by burning sulphur in rooms where many cases have occurred. His observations show that the germs actually float in the air, in quantities sufficient to be recognized under the microscope, and only awaiting a favorable "soil," to germinate anew. With regard to the often-asked question, "When is the child fit to go back to school?" Dr. Fox says, if the hair is dull and dry, suspicion should be excited, and then adds this practical rule—" Condemn a child any portion of whose scalp is studded with the little dark points of short broken-off hairs." In any doubtful case the microscope is necessary, and it is safer to err on the overcautious side, in cases where the root seems healthy, but the shaft covered with epithelial and exudation matters at its follicular portion above the root, albeit this *may* be but the result of the irritation of remedies.

The *treatment* of these diseases will comprise the destruction of the fungus; this is to be effected by the use of the iodide of sulphur ointment or a strong solution of nitrate of silver, or a strong lotion of sulphurous acid, or the mixed vapors of iodine and sulphur may be applied to the part. The vapor evaporated from gr. iv. of iodine and ℥ j. of sulphur applied twice or three times a day, or ointment of creosote, or white precipitate gr. xx. to sulphur ointment ℥ iv., or lastly, the liquor epispasticus. Perhaps of these the iodide of sulphur and the white precipitate and sulphur ointments are the best. Cleanliness is essential; the scalp should be frequently washed and bathed. Tonics are often valuable, as steel and cod-liver oil, the value of which last medicine, properly assimilated, Dr. T. Fox attaches especial value to. The disease, though it persist long, does not cause baldness.

Tinea favosa attacks not only the scalp, but also the chin, eyebrows, and forehead, and sometimes even the trunk and extremities. It consists of small cup-shaped yellow crusts, dry, and looking something like a honey comb; each cup has a hair in its centre. The odor emitted is very offensive; it has been compared to that of a mouse. It appears about the age of seven. In England it is a rare disease.

The parasite causing the disease is called *Achorion Schönleinii*. This disease, if not arrested, destroys the follicle, and permanent baldness is the result. Sir W. Jenner has pointed out that herpes circinatus is as favorable to the growth of *Achorion Schönleinii* as it is to *Trichophyton tonsurans*. It is contagious.

Treatment.—Constitutionally, as in the preceding disease. Locally a lotion of Hyd. Bichlor. (gr. vj., Aq. ℥ j.), or Cupri Acetat ℥ ss., Adipis ℥ j., are good applications; or a strong solution of sulphurous acid may be applied to the part. If on the hairy scalp, epilation has often to be resorted to, and the hairs should be extracted in the direction in which they grow. Professor Cantoni has recently advocated the use of rectified spirit for the destruction of the Achorion after the crusts have been removed by poultices.

Tinea decalvans (*Alopecia circumscripta* or Porrigo decalvans). This consists of smooth bald patches, without pain, heat, or redness. The roots of the hair have atrophied, until, being smaller than the follicles, they fall out. The parasitic fungus in this case is *Microsporon Audouini*. It is contagious, and common in children especially between the ages of five to ten years. Dr. Duhring, of Philadelphia, denies the existence of the *Microsporon Audouini*, and says that the appearances commonly referred to a fungus are, in reality, merely a débris of epidermic scales—sebum—which adheres closely to the hair on being treated with a reagent.

The constitutional treatment is as in the other parasitic diseases. Locally Tinct. Iodi applied night and morning, or a strong solution of sulphurous acid, or Liquor Epispasticus in very obstinate cases. The baldness is never permanent, and the first sign of improvement is the appearance of soft downy hair on the bald patch.

T. sycosis (Mentagra) does not occur in children.

Chloasma (Pityriasis versicolor), or liver spot, appears usually on the front of the chest and abdomen. The patches are of a dull brownish-yellow color. It may last for years, and is contagious. The parasitic plant is the *Microsporon furfur*. Want of cleanliness favors its production, according to some writers—especially the wearing of flannel waistcoats seldom washed; but the late Mr. Startin saw many cases where every attention to cleanliness had not prevented the disease. It is especially liable to occur in those who perspire very freely. It has been observed in children, but is not common. In treatment cleanliness and frequent washings are most important. The sulphurous acid lotion or Hyd. Bichlor. (gr. iij.— ℥ j.) may be used. Mr. Startin considered a course of arsenic essential to the permanent cure.

Scabies or Itch is a vesicular disease, caused by the *Acarus scabiei*, a small animal parasite, which is found about a line from such vesicles. It mainly attacks the hands between the fingers, but no part is exempt, though the face is rarely attacked. After the disease has lasted some time, cracks appear, and excoriations, owing to the scratching the itching of the disease occasions.

It occurs on the soles of the feet in young children, not unfrequently. Scabies is often complicated with eczema, ecthyma, and other cutaneous affections to which this irritation gives rise.

Sulphur ointment is the general remedy for this disease. The patient, having been thoroughly washed, should be rubbed over with it wherever spots exist. If there be unusual thickness of the cuticle, the ointment may be

℞   Adipis... ℥j.
    Sulphur. Præcip..................................... ʒij.
    Pot. Bicarb... ʒj.
Ft. Unguent.

A strong alcoholic solution of stavesacre is also efficacious, or the Pulv. Staphisagriæ may be combined with Ung. Sulph. The oil of chamomile is also stated to be useful in the Ung. Sulph. A lotion of pentasulphide of calcium is recommended by Mr. Erasmus Wilson. If it be desired to conceal the fact of using sulphur ointment, it may be colored with Hyd. Bisulph. (Cinnabar), and scented with oil of bergamot. Styrax is also occasionally employed, e. g.,

℞   Styracis Liquid..................................... ℥j.
    Sp. Rectif.. ʒij.
    Ol. Olivæ... ʒj.
Ft. Unguent.

CHAPTER IV.

CONGENITAL AFFECTIONS AND DISEASES OF THE NEWBORN.

UNDER this heading it is usual to group together several affections, some of which, however, will be treated of in those chapters to which they seem more properly to belong. The list comprises Asphyxia neonatorum, Cephalhæmatoma, Diseases of the navel, Trismus, Sclerema, Ophthalmia neonatorum, Cyanosis, Hydrorachis, Icterus, Atelectasis pulmonum.

Asphyxia neonatorum is that condition in which on birth the child does not breathe; the child is alive, as evidenced by the beating of the heart, but the inspiratory muscles do not act. The more correct name for the disease is, as Dr. Tanner has pointed out, "Apnœa." The common causes of this condition are long-protracted labor, the cord being twisted round the neck of the infant; great compression of the cord from prolapse or other causes; the mouth and fauces being choked up with viscid mucoid secretions; and lastly, extreme weakness and debility of the infant. Very simple measures are usually sufficient to overcome this condition. Carefully wiping out the mouth and throat, and cleansing it from the mucus with which it is plugged, and slapping the child on the buttocks. Desormeaux recommends that a little spirit be held a moment or two in the mouth, and then spirted with force on the breast of the child. Wiping out the mouth often causes choking and coughing, and so sets respiration going; or, if these means fail, a warm bath may be employed and a dash of cold water used immediately afterwards and the child again popped into the warm bath. The alternate heat and cold, especially with friction, often has the desired result; when respiration commences the child should be at once removed from the bath, too long immersion in which will only do mischief. A cold douche on the stomach, recommended by Dr. Tott, and inflation with air combined with gentle movements of the chest, as used in artificial respiration, are means to which it will be seldom needful to resort. Electricity, when at hand, might be employed in an extreme case. In apnœa the cord is not to be cut too soon, except the case be cyanotic, or evidently congestive or apoplectic in origin, when a tablespoonful or so of blood may be allowed to flow with advantage before the cord is tied. Dr. Maschka has shown that efforts are not to be relinquished hastily—an infant may be an *hour* or more without breathing and still be saved.

1. DISEASES OF THE NAVEL.

The navel should be wrapped up in a piece of soft linen rag, in which a hole has been made for it to come through, and laid gently upwards on the abdomen; the flannel band over keeps it in its place. Its detachment is usually complete in five days or less, those cords soonest coming off which are dry and thin and contain less Whartonian gelatinous matter. When the separation is effected favorably there is merely a little oozing of serum, and by the end of a fortnight the cicatrix is healed. Sometimes, however, things do not run so satisfactory a course, inflammation may take place and a sero-purulent discharge be poured out, and erysipelas of the surrounding textures may occur. Slight inflammation or ulceration will require, especially, attention to cleanliness, and a little weak lead lotion or zinc ointment may be applied. Now and then a touch with caustic will be useful, especially if the wound have an indolent character. If any tendency to sphacelation come on, disinfectant dressings, carbolic acid lotion, or a lotion of Condy's fluid must be combined with careful support at the breast, aided by a little wine or brandy in milk or coffee.

Hæmorrhage from the navel is generally sufficiently serious. It is fortunately rare; its occurrence varies from within a few hours to the eighteenth or twentieth day after birth, usually about the eighth day. Icterus is not an uncommon prodroma of this hæmorrhage. The blood is usually very slow to coagulate, and appears as a continuous oozing rather than as a stream. Collodion and plaster of Paris are the most hopeful dressings to stop this hæmorrhage; ordinary styptics are of little service, and operative measures, especially transfixing with pins, &c., seem merely to excite great bleeding from the newly wounded points. It must be remembered that a hæmorrhagic diathesis underlies this condition, and that it is often associated with diseased liver and constipation. The mineral acids and other astringents internally are remedies from which good is often derived in very marked manner. In other cases free purgation followed by astringents has been successful. I have recently attended a woman who has lost three children from this cause. She has suffered from secondaries, and this pregnancy submitted to antisyphilitic treatment. The hæmorrhage from the navel, however, occurred as after the former confinements, and styptic colloid, plaster of Paris, and all other means failed to arrest it.

2. Sclerema.

Sclerema, or induration of the cellular tissue, with or without œdema, is a disease almost exclusively confined to the first weeks of life. It usually commences at the feet or calves, sometimes about the pubic region, and spreads over the greater part or, perhaps, the whole of the body except the thorax, which is almost always, if not always, exempt. The swollen parts are hard, the skin tense, so that it cannot be pinched up or moved over the parts beneath; it is dry, harsh, and cold to the touch, and may vary in tint from a yellowish to a purple hue. According to Léger a thermometer in the mouth sinks to 73° F.; every vital function is depressed; the breathing slow, the voice and cry weak ; sucking feebly performed ; the pulse falls, according to Valleix, often to sixty beats per minute. The sensibility of the skin is lost. After a time cough comes on and continues throughout and indicates the occurrence of either lobular pneumonia or atelectasis pulmonum, which last is by some considered the cause of sclerema. There can be no doubt of the frequency of association between the two conditions. It seems most probable that there are two distinct varieties of sclerema, one due to serous infiltration of the cellular tissue, and the other to induration of the adipose tissue. It is a very rare disease in any form.

The general symptoms accompanying the "skin-binding" are pain, indicated by a sharp, shrill often repeated cry, convulsive movements, failing strength, constipation, dysphagia, scanty urine, a gradually increasing condition of stupor. Death usually occurs on the fourth day, though it may be later. In cases which recover, according to Valleix, the eyelids and forearms are the first parts to become flaccid, the legs and feet remaining longest œdematous, and so long as they do so remain the patient cannot be considered out of danger. The affected parts retain their discoloration after the swelling disappears, and the skin is also weak, flabby, and crinkled for some time. Of fifty-three cases reported by Elsässer only four recovered.

The treatment comprises first and most essentially warmth, warm baths, hot water or vapor baths, frictions, and a good temperature of the room in which the child is. Stimulants are also to be given, and the child should have the breast if it can take it, and if not the milk should be milked out of the breast and given with a spoon. Richter, Eberle, and Dr. Condie speak highly of a blister applied early to relieve the engorgement and to promote the absorption of serum. Dr. Condie recommends that it be left on three hours and succeeded by a large emollient poultice. Valleix records two recoveries in which two leeches were placed behind each ear; but in other cases, under the same physician, this plan was not successful.

Punctures to relieve the tense condition and give exit to the accumulated serum seems rational, and would probably be attended with benefit.

3. OPHTHALMIA (NEONATORUM).

This disease usually appears within a few days of birth. It is seldom so long as a week in showing itself. The eyelids are found one morning to be sticking together, and there is also some little redness and swelling. On opening the lids the conjunctiva is seen to be very inflamed and covered with a sticky transparent coating. Some purulent matter forms, the lids get more agglutinated, and there is more swelling. There is extreme intolerance of light. The cornea may become hazy, and sometimes purulent infiltration or ulceration ensue followed by prolapse of the iris; the sight is, of course, then lost. Complete opacity of part or the whole of the cornea from interstitial deposit is not uncommon, a thin film forming over its surface. Besides those cases attributable to external and direct irritants, as exposure to too strong a light, or the dropping of irritating substances (soap, *e.g.*) into the eye, defective hygienic conditions, especially such as are consequent on the presence of puerperal fever, are specially favorable to the fostering of the disease. A leucorrhœal and notably a gonorrhœal discharge in the mother is also a particularly common cause. The prognosis must be formed on the state of the cornea, the earlier it is implicated the graver the case. The amount of œdema of the lids is generally pretty well proportioned to the severity of the case. Opacity of the cornea from mere thickening of the conjunctival covering may disappear with time, but that from ulceration is permanent and interferes in proportion to its extent with sight.

Treatment.—As struggling is probable and dangerous, the child should be rolled in a shawl, its arms being close to its sides, leaving only the head free; small pieces of rag will then be useful in cleansing away the discharge; these should be instantly burned when done with, or a small sponge, if kept absolutely clean, may be employed. Syringing carefully under both lids with tepid water will complete the cleansing process, which must be thoroughly performed. Next a solution of alum (gr. vj. ad ℥ j.) should be injected, and care taken that it is diffused all over the cornea. The lids should be dabbed with soft rag to dry them, and a little fresh lard or spermaceti ointment smeared along the edges to prevent adhesion. This process is to be gone through frequently; every two or three hours will not be too often at first, and afterwards four or six times a day. M. Liebreich advocates the use of mitigated nitrate of silver, that is, a mixture of one part nitrate of silver, and two parts nitrate of potash, melted and poured into an iron mould. "The eyelids," he says, "must be reversed one after the other, and after being carefully cleansed touched with this caustic, which must be passed all over the swollen and red part of the mucous membrane; before replacing the lids the free nitrate of silver should be neutralized by a drop of salt water." One application of caustic a day is sufficient, and as soon as improvement occurs once every two or three days. This treatment cannot be employed at all unless the doctor can see the child daily. If ulceration threatens, a solution of atropine should be dropped into the eye to dilate the pupil and allay inflammation (F. 212), and quinine, or bark, or syrup of the phosphate of iron administered. Sometimes, when inflammatory symptoms have all passed away, the conjunctiva remains relaxed and thickened (granular conjunctiva), a mustard plaster on the temple, a Collyrium of Vinum Opii, and tonics will suffice to relieve it.

CHAPTER V.

FEVERS.

1. MEASLES—RUBEOLA—MORBILLI.

AN acute specific disease—febrile and infectious, ushered in with catarrhal symptoms and characterized by an eruption of the skin, which appears usually upon the fourth day.

Usual symptoms.—After a period of incubation varying from twelve to fourteen days,* there is manifested alternate chilliness and heat, a quickened pulse, aching in the limbs, slight headache, soon followed by redness of the eyes, coryza, huskiness and hoarse cough. On the fourth day there is an eruption of soft, circular, very slightly elevated dusky red spots, which appear first on the forehead, and extend over the face, neck, and whole body. The spots gradually coalesce and present a peculiar crescentic or horse-shoe shape. The spots disappear on pressure. They attain their greatest intensity on the fourth day from their invasion, and by the seventh day they fade away with a slight desquamation of the cuticle. As a rule the fever does not abate on the appearance of the eruption.

M. Girard thinks the contagion of measles is active during the prodromic stage. He also states that red spots are visible on the velum pelati four, five, or six days before the eruption appears on the skin. Dr. Ringer states that the highest temperature reached in ordinary measles is 103° F.; the decline of temperature is sometimes as early as the fourth day, sometimes as late as the eighth or tenth day.

Occasional symptoms.—There may be no prodromata whatever, or the attack may be ushered in with convulsions (especially in children), or there may be delirium, or there may be a great amount of fever, or there may be and often is sore throat; more rarely severe headache, and sometimes absence of the coryza.

The eruption may be scanty, or most abundant and confluent, but the quantity of the eruption *per se* does not affect the gravity of the attack; the color of the eruption may be dark, constituting so-called "black-measles;" there may be petechiæ, which do not fade on pressure and resemble purpura; these do not *per se* affect the prognosis. Miliary vesicles are often present, and when abundant the amount of desquamation will be greater.

Complications.—Rilliet and Barthez point out that measles manifests itself by a double inflammation, that of the skin and that of the mucous membranes. That which is upon the skin should predominate, if not the mucous membranes will suffer unduly, and pulmonary, laryngeal, or intestinal inflammations result. The common complications are accordingly found to be bronchitis (very common). Collapse of a portion of the lung. Pneumonia—this should be especially looked for at the apex of

---

* The period of incubation in cases produced by inoculation is seven days.

the inferior lobe; it is often insidious and unsuspected. Laryngitis, croup, and otitis, the last common towards the close of the disease. Ophthalmia.

Of 167 cases recorded by MM. Rilliet and Barthez, bronchitis occurred in 24, pneumonia in 7, lobular broncho-pneumonia in 58, and laryngitis in 37. The initial stage is that in which the complications most frequently appear. A pulmonary complication occurring in the prodromic stage often retards the eruption or renders it irregular or imperfect, in the second stage it may cause actual retrogression. Extra drowsiness and a quick full pulse are often the most characteristic symptoms pointing to coming pulmonary mischief, which should then be especially watched for.

Sequelæ.—Diarrhœa—this in moderation is beneficial, and should not be interfered with. Albuminuria and dropsy (rare). Tubercular deposits. Pertussis very common. Parotitis.

Average mortality.—One in fifteen.

Prognosis.—If uncomplicated favorable. Unfavorable signs are great fever, great dyspnœa, sudden vanishing of the rash, together with an access of delirium; brown dry tongue, with special severity of some two or three symptoms; petechiæ, with a typhoid form of fever. Capillary bronchitis and pneumonia are the most frequent proximate causes of death.

Treatment.—The child must be kept in bed in a large well-ventilated room, free from draughts—a point of vital importance, looking to the frequency and danger of chest complications. The diet must be low. Tepid drinks may be freely given. It is very important in measles as in all infectious fevers to remove all discharge and soiled linen instantly; the motions should be passed into vessels containing chloride of lime, carbolic acid, or Condy's fluid; this with ventilation will go far to prevent infection. There is no objection, if it be grateful to the patient, to have the body gently sponged with warm water; and if itching be much complained of inunction with unsalted lard is useful. Cough is often the first troublesome symptom which requires special treatment. A mixture containing citrate of potash and ipecacuanha wine, with a few drops of nepenthe or Tinct. Camph. Co., will usually quiet this. If the fever runs high, the weak mineral acids sweetened and largely diluted will be very grateful. Or a mixture of citrate of potash and Rochelle salt may be given in an effervescing form. If the fever be of a low type, with brown tongue and failing powers, large doses of chlorate of potash will be useful, and stimulants will be required. Yolk of egg beaten up with wine is excellent in such cases. Purgatives, as a rule, are not required; if employed they should be mere laxatives, remembering the diarrhœa, which usually sets in towards the close of the disease. In cases attended with much nervous excitability and convulsions or delirium, bromide of potassium in full doses will be useful. This drug will also procure sleep and is better for the purpose than any opiate. Sudden recession of the rash attended with an onset of delirium should be met by plunging the child into a bath containing mustard, and leaving it in until the surface becomes red, which usually occurs in a few minutes. The child should then be rolled in a blanket, and the strength supported by nutritious diet, and stimulants as needed. For laryngitis, a sponge wrung out of very hot water should be applied over the larynx, and inhalation of steam encouraged. Pneumonia will call for a stimulating embrocation over its site, and the administration of stimulant expectorants—carbonate of ammonia with senega is the best.

Lung and indeed all complications occurring during the early stages are best treated by endeavoring with external stimulants, *e.g.*, the mustard bath and internal gentle diaphoretics, to get the rash thrown out freely. Later on this is, of course, inadmissible, and the strength must be supported in every way. Ipecacuanha seems to stand out as the drug most generally serviceable in the complications of measles; it checks diarrhœa, it loosens phlegm, and it promotes diaphoresis, all valuable qualities. When pain is present, especially if at all severe, Dover's powder will replace simple ipecacuanha powder with advantage.

As the disease declines the diet may be more solid, and tonics will be of service. Convalescence from measles is often slow, and as discharges from the ears, eyes, and nose are not uncommon, sea air is very beneficial in re-establishing the health. Such discharges will require astringent lotions and the use of cod-liver oil and steel.

2. Rötheln (Rubeola notha).

Under this name is described a disease which partakes in a measure of the characters of both measles and scarlet fever, and is yet distinct from either, and especially in this, that attacks of rötheln confer no immunity against subsequent attacks of either scarlet fever or measles and *vice versâ;* in fact, cases are on record where an attack of rötheln has been followed almost immediately by an attack of genuine rubeola or scarlatina. The chief points about rötheln are as follows: the prodromata are slight, or even wanting; the rash which at first closely resembles that of measles is of *very short* duration, disappearing in a day, rarely indeed lasting two days. Catarrhal symptoms are slight or absent and desquamation is very partial and incomplete. The temperature, in my own experience, has been remarkably low, rarely reaching and never exceeding 100° F., sometimes ranging from 97° F. to 99° F. The child is not nearly so generally ill as with scarlatina or even measles, though for a few hours a sharp attack may seem impending; the tongue is comparatively little furred and presents no "strawberry" appearance. There may be some angina, and the rash may, especially after its first appearance, assimilate a good deal to that of scarlatina, though it is always more confined to patches. The abdomen is the part in my own experience most covered, but others speak of the face and neck and upper extremities as being most prominently affected. The color of the rash is something between the scarlet of scarlatina and the lake-like color of measles; it is rather elevated, and the patches are of most irregular shape. The disease is certainly far less contagious than scarlatina, but it undoubtedly is contagious. In the Victoria Hospital rötheln appeared in a child who was in a ward containing four beds—two opposite the other two. The child in the bed on the same side of the ward took rötheln in well-marked form, the other two children in the opposite beds escaped. It is so mild and unimportant a disease that the simplest febrile treatment is sufficient. There are rarely complications and no sequelæ.

3. Scarlet Fever—Scarlatina.

An acute specific disease—febrile, contagious, and infectious, and accompanied by a peculiar eruption of the skin. After a period of incubation varying according to different authors at from four to forty days, and probably averaging from four to six days, there appears in children vomiting; in older persons sore throat, and the onset is usually sudden. It is common for adults to be able to fix the hour in which the sore throat began. In children severe vomiting often prognosticates severe throat affection. Next there is noticed fever, a frequent pulse, commonly 130 to 170, a flushed face, a high temperature (103° or 104° F. even on the first day), hurried breathing, furred tongue, hot skin and thirst. At the same time there is lassitude and restlessness, headache, and at night delirium. On the second day usually about the root of the neck and upper part of the chest, appears the eruption, which is a scarlet efflorescence consisting of innumerable red spots at first separated by natural skin, but soon coalescing and producing a general redness; the skin is rendered pale by pressure, but the redness immediately returns—the rash is not elevated to the touch. It is most abundant about the hip and loins, and the flexures of the joints, in fact where the papillæ of the skin are largest. The eruption reaches its maximum intensity on the third or fourth day; by the fifth it has begun to fade, and by the eighth it disappears. It goes off in an order corresponding with its invasion. Miliaria are often present, perhaps more commonly in adults than in children; they in no wise affect the prognosis. The sore throat is very important, especially in children. A child may die from throat disease without any complaint about its throat having been made. The throat should therefore always be carefully examined. The tonsils will usually be found enlarged and inflamed, and often coated with a thick white tenacious mucus. Superficial ulceration of the tonsils is sometimes present, but ulceration of any other portion of the mouth or throat is uncommon, except in the malignant form of the disease. Œdema often occurs, and the glands at the angle of the jaw become tender and swollen. The papillæ of the tongue are elevated and project as bright red points through the white mucus on the surface, or the whole tongue may be vividly red with prominent papillæ. These conditions produce the well-known and highly characteristic white and red "strawberry tongue." The fever does not, as in variola, abate on the appearance of the rash, but declines with it; the temperature may reach 106° Fahr., and is usually at its height by the fifth day. There is generally an increase in the amount of fever towards night and remission towards morning.

Dr. Ringer's observations go to show that in the most severe cases the temperature remains equally high all day, in less severe cases it remits a little in the morning and rises to a height between 2 and 8 p.m. When the morning remission is well marked it is a favorable prognostic. The fifth, tenth, and fifteenth days are those usually marked by a decided fall, and if after such decided fall the temperature again rises it indicates the development of some sequela, either an inflammation of the kidneys, throat or one of the serous membranes.

The pulse continues frequent, varying from 120—160, is quicker in children than in adults, and declines gradually with the fever. The urine, as in all fevers, is scanty and high colored; during the decline of the disease it may contain albumen, as will be more fully noticed presently.

The urea is not increased, the chlorides are diminished. The phosphoric acid after the first three or four days diminishes even down to one-half or one-third its normal amount (Gee). Uric acid is retained during the fever and voided in excess on its decline. Bile pigment is present during the first six days.

The bowels are generally confined. As the rash fades desquamation of the cuticle commences in the order of invasion; the desquamation in delicate skins is furfuraceous, but from the hands and feet it is generally in scales; sometimes pieces like portions of a glove are thrown off from the hand. The desquamation may last from a few days to a week or more. If miliary vesicles have been present it is generally earlier and more complete. It is attended with a good deal of itching, irritation, and tenderness. The disease is considered especially contagious during desquamation.

The duration of an ordinary case of scarlet fever is about a fortnight.

Varieties and deviations from the type—S. maligna.—The tonsils in this form become the seats of sloughing ulcers, leaving ragged sores, but the especial character is the pyrexia, which is of the so-called malignant type; the excitement great, the delirium violent, followed by extreme ataxia, and exhaustion. Sometimes the excitement period is so short and so soon followed by the depressed or typhoid condition that the disease may kill within forty-eight hours of its invasion.

S. anginosa.—In this form the stress falls upon the throat. The swelling of the glands and cellular tissue around the neck is so great as to form a so-called "collar of brawn" from ear to ear. There may be ulceration and sloughing of the fauces and pharynx, and posterior pharyngeal abscess. The mouth is opened with difficulty, there is great dysphagia, and liquids return through the nose. The inflammation often extends to the ears through the Eustachian tube, and is followed by a purulent discharge from them.

The rash in these deviations from the type is of little clinical significance; it is often abundant and of good color. In the worst cases death occurs before the rash appears.

Scarlatina sine Scarlatina or S. latens.—This form occurs oftenest in those who have had an attack of scarlatina, and become exposed again to its contagion. But it is important to remember that the mildest form of scarlatina may give rise to the very gravest sequelæ, and to the most malignant varieties in persons exposed to their infection.

Sequelæ.—For the sake of convenience some affections already mentioned will be rearranged here.

1. Ulceration and sloughing of fauces and pharynx.
2. Retro-pharyngeal abscess.
3. Scarlatina bubo, which may be either an inflammation and suppuration of cellular tissue around the parotid, that gland remaining itself unaffected; or inflammation and suppuration of the lymphatic glands around the parotid, the parotid still unaffected; or more rarely, inflammation of the parotid itself.
4. Bronchitis and pneumonia rare, pleurisy and pericarditis more common.
5. Otorrhœa, sometimes followed by permanent deafness.
6. Diarrhœa, a dangerous sequela often depending on follicular enterocolitis or softening of the intestinal mucous membrane.
7. Abscesses.
8. Joint affections.

Tenderness, redness, and swelling of the joints are all very common after scarlet fever. These symptoms mostly occur about or a little before the time of desquamation; moreover a tendency to suppuration is shown, which in true rheumatism never appears. About the same time is often heard a systolic murmur at the heart's apex, but this is by no means necessarily due to rheumatic endocarditis, as it may arise without any quasi-rheumatic pains being present; this murmur may persist for a month or more—may remain constant or may gradually die away.

9. Renal dropsy (renal catarrh, tubal nephritis).

This is a common and very fatal sequela. It occurs in about one-sixth of the cases. Albuminuria is usually the first symptom of the coming mischief. It may occur as early as the first week, or it may not be found till the third. The twenty-second day has been found to be a common day for dropsy to make its appearance. With the albuminuria is considerable pyrexia—the skin becomes again hot and dry—the process of desquamation ceases, the appetite is lost, the bowels become constipated, and in a day or two puffiness is noticed about the eyes, followed by some œdema of the hands and feet, and these again may be followed by large serous effusions into the various serous cavities, the swelling and anasarca may increase, the urine becomes more and more scanty. If the urine be examined it is found to be smoky in appearance and highly albuminous. The microscope reveals blood-corpuscles, epithelial cells, and granular and epithelial casts.

The relative quantities of albumen and blood sustain no relation to each other. Albumen may be abnormal and blood-corpuscles absent, or large quantities of blood may be present, and only a trace of albumen. Dr. Basham points out that a bluish-green, tending to black, reaction with nitric acid and heat, is a grave prognostic, and indicates extensive renal disease.

The final effusion may be peritoneal, which is the least fatal—pericardial, which is known by sudden and urgent dyspnœa—blueness of the face and fluttering pulse—a tendency to faintness—besides the ordinary physical signs of that condition; and lastly, hydrothorax, which is recognized by the sudden accession of vomiting, dyspnœa, and lividity, accompanied with great restlessness, anxiety, suppression of urine, and death within twenty-four hours. In mild cases, which go on to convalescence, albuminuria is often persistent for a long time after the disappearance of the dropsy. A chill during desquamation is the common cause, and the first signs of amendment are the disappearance of the dropsy and copious diuresis. During convalescence, a child will pass from three to five or six pints of urine in twenty-four hours.

10. Scarlatinal vaginitis, or muco-purulent discharge from the vagina, is by no means rare, but is often passed over, unless so severe as to compel attention. I have seen several very obstinate cases of persistence of this discharge, one in particular which lasted eighteen months, and was accompanied by severe pain in micturition. This case, after a great variety of treatment under different practitioners, finally yielded to two or three applications of nitrate of silver about the orifice of the urethra (where there were a few very fine projecting red points), an injection of oak bark, and quinine and steel in good doses.

11. Lastly, diphtheria, or one or other of the acute specific diseases, may supervene in the course of scarlet fever.

Post-mortem.—Scarlet fever possesses no special anatomical character. The lesions after death will be those of the special complication by which death was caused.

Law of Infection.—The disease is communicated from person to person, and by clothes, and its infecting power through the air is also considerable. It is usually a non-recurring disease. If it recur, it is mostly in a mild form, *e.g.*, S. latens. It is often epidemic, and different epidemics present marked differences in many of the more prominent characters, and especially in the tendency to acute desquamative nephritis.

Prognosis.—It is probable that pregnant women are peculiarly exempt from the liability to take scarlet fever; but the puerperal state is one peculiarly predisposing to its reception, and in such cases it is extremely fatal. Unfavorable symptoms in the ordinary course of the disease are a tendency to typhoid or malignant type; early delirium, especially if accompanied with vomiting and hiccough; convulsions and coma; parotid abscess; bronchitis; pneumonia; albuminuria; anasarca; serous effusions; suppurative arthritis.

Prophylaxis.—The temperature of 212° F. destroys the morbific principle; hence the clothes &c., may be baked. Belladonna and other reputed specifics are in reality useless. To prevent infection clinging to a room, carpets, curtains, and hangings should be dispensed with. Fresh air must be freely admitted; all secretions are to be passed into vessels containing carbolic acid or chloride of lime, and instantly removed. The surface may be sponged with weak Condy's fluid, basins of which may stand about the room. By rigorous observances of these rules the disease may be prevented from spreading through a house.

Dr. Brakenbridge and Dr. W. Scott have recently called attention to the value of sulpho-carbolate of sodium in doses of from 5 to 30 grains according to age, and given three or four times a day as a prophylactic not only in scarlatina, but also in diphtheria and measles. It was given in 7 families to 22 persons exposed to scarlatina, in 3 families to 15 persons exposed to diphtheria, and in 3 families to 8 persons exposed to measles. In every instance the contagion did not spread beyond the individual first attacked. The rationale of the sulpho-carbolate is the disinfecting the tissues of the body and so removing any favorable nidus for the development of fever germs. The matter well deserves further investigation.

Treatment.—If the general and local diseases are alike mild, watching and nursing will be the chief elements of the treatment. The child should be put to bed in a large well-ventilated room. It should not be too heavily covered with clothes. The blankets, sheets, and body linen should be frequently changed. A fire should be burnt in the room, and the windows left open, draughts being, of course, avoided. The whole body may be greased with unsalted lard, night and morning. This greatly relieves itching and irritation. The surface may be sponged with tepid water or Condy's fluid and water, but chill must on no account be permitted. The diet must be plain and simple, without stimulants. An infant at the breast should be kept to the breast; older children may have milk and water, light puddings, and gruel; lemonade, toast-and-water, and other drinks may be freely allowed. It is better not to open the bowels by drugs, unless such are really needed, and then the aperients should be gentle.

The best local application to the throat is a warm fomentation, *e.g.*, a linseed-meal poultice, a yeast or oatmeal poultice, a poppy stupe, frequently changed. Inhalation of steam is also useful. The best medicines are refrigerants, such as chlorate of potash or sulphuric acid, in rose infusions, sweetened. Large doses of chlorate of potash may be given to the youngest infants in the adynamic type of fever. The throat may be sponged or syringed out with a gargle of chlorate of potash and honey, or tincture of kino, or chlorinated soda, or the Inf. Rosæ Acidum (B. Ph.). These are all useful in removing the tenacious ropy mucus which is so great a torment to many patients.

The use of veratrum viride is highly recommended by American physicians, the drug exercising, as in acute inflammation, a reduction of the pulse, a subduing of undue nervous erythism, a diaphoresis of the skin, without, as is affirmed, any permanent prostration resulting. The throat affection is also stated to be improved under its use, and the tendency to sequelæ diminished. Aconite may be similarly used (see Aconite in the

Formulary). Sulpho-carbolate of sodium is recommended by Dr. Sansum in five to ten grain doses.

Dr. Sweeting has quite recently recommended ammonia and milk; he gives it in this form:

℞ Spir. Ammon. Arom... ʒiij.
   Sodæ Pot. Tart. .... ʒj.
   Tinct. Lavand. co.. ʒj.
   Aquæ, ad.. ℥iss.
Ft. Mist.
      Coch. j. min., tertia quaque horâ. For a child 3—6 years old.

Also undiluted sweetened milk as diet.

Dr. Sweeting condemns all external treatment, as sponging with vinegar and water, the cold douche, &c. Dr. Eddison and others speak highly, on the other hand, of baths commencing at 90° or 100° and cooled down slowly to 70°, where the fever runs high and the temperature is excessive. Danger from "driving in the rash" is asserted to be without foundation. There is generally a rapid rise in temperature in removing the child from the cooled bath, and the expedient has to be resorted to again, except Dr. Eddison's plan of keeping the child in "several hours" "even for a day or longer" be adopted. In judiciously selected cases where hyperpyrexia is present a cautious use of the "cooled bath" is no doubt of benefit, but I can by no means advise its indiscriminate use, nor its employment for extended periods. Tepid and cold sponging will often be found sufficient, or allowing the patient to paddle the hands in basins of gradually cooled water (see Typhoid Fever).

Mr. Taylor, of Liverpool, speaks highly of the wet sheet; he slits up a nightgown in front and immerses it in hot water, either pure or medicated with a drachm or two of tincture of capsicum, or a little mustard. The "sheet" should be well wrung out and applied suddenly, the patient being then packed in two blankets; a pillow or two or an eiderdown quilt to be thrown over all. Mr. Taylor warmly advocates the "medicated" sheet in many affections besides scarlatina, and I think we may all agree with him thus far "that persevering cutaneous elimination is a great medical power." My own feeling is this, we have many and excellent means to our hands none of them to be used empirically; all of them may be of value in special cases. An ordinary case of scarlatina will run its course perhaps as well without as with remedies. But the prudent physician will remember he holds strong means in reserve for all accidents; aconite, for hyperpyrexia, or "the cooled bath"; chlorate of potash for throat mischief; and if the pulse be low, the skin pallid, cerebral symptoms impending with delirium, and other signs of evil omen, then the packing with hot mustard and water may have the best effect. Let us endeavor to remember all reasonable means of giving relief, that our armory may not be without a weapon in the hour of need.

A regurgitation of fluids through the nose points to the existence of ulcerations, or more often to retro-pharnygeal abscess. This should accordingly be searched for and opened by cutting through the pharynx towards the spinal column, as it often burrows long and seriously before opening of itself. Ulcerative patches should be well cauterized with the solid nitrate of silver. In the adynamic type carbolic acid in small doses is well spoken of. The occurrence of otorrhœa and suppurative discharge of the nostrils will require quinine and syringing—the ear with warm water, the nose with a weak solution of sulphate of zinc or nitrate of silver. All intercurrent possibilities are to be narrowly watched for, the chest auscultated from day to day, the urine examined, the temperature taken. Rheumatic pains will require that the affected joints be wrapped in cotton wool; and the administration of small doses of iodide of potassium will be attended with the best results, whether mitral murmur be present or not. Such coexistence may, however, in addition, demand an occasional application of blistering fluid, or even a few leeches; but such cases are rare, as the constitutional depression usually far outweighs in importance all local imflammations; and this must be borne in mind, whether the inflammation be pneumonia, bronchitis, or whatsoever stimulants are constantly needed, and the sesquicarbonate of ammonia with senega, the most appropriate drugs when cough is troublesome and expectoration deficient. Ice is of value, and may be fearlessly given to allay cough or vomiting, which are often concomitants of albuminuria. For dropsy, elaterium is the drug commonly recommended; its tendency to excite vomiting I consider a great objection to its use. When employed, the dose is one-twelfth to one-sixth of a grain for a child nine years old, repeated every three or four hours. I prefer compound jalap and scammony powder, with or without a little nitre or calomel, as need requires. About twenty grains should be given to a child ten years old every four hours, till copious diuresis and purgation result. Counter-irritation over the loins by sinapisms or turpentine stupes is useful, so are warm baths. After a few days Tinct. Ferri Perchlor., with or without quinine, will be needed.

Convalescence from scarlet fever is a time for the exercise of much firmness and patience. The child must be kept at home when there is apparently but little the matter with it, both for its own sake and also for that of others, the disease being undoubtedly contagious during the whole period of desquamation. Convalescents must be warmly clad, wear flannel next to the skin, have generous diet, and, after a month or six weeks, get the benefit of seaside or country air.

4. Typhoid Fever.

An acute specific disease, slightly infectious and contagious, associated with a peculiar eruption on the skin, and disease of the solitary and agminated glands of the intestines. The period of incubation is about a fortnight.

This disease is the so-called infantile remittent fever, bilious fever, gastric fever, and mesenteric fever, of different authors.

Symptoms.—Sometimes the disease is so slight, and runs so quiet a course, that but two recognizable symptoms manifest themselves; but these are most significant, viz., loss of muscular power and heat of skin, as shown by a thermometer under the axilla or tongue; and it is important to be aware that a disease manifesting but these two symptoms, with some little poorliness and general malaise in addition, even in a healthy young adult, may prove fatal suddenly in the third or fourth week, and that such a case will require to be treated with the same care and precaution that we should give to the form manifesting any or all the more alarming symptoms we have now to discuss. The first in importance, and one often earliest observed, is diarrhœa: it may be slight or severe—sometimes so slight that the patient or the patient's mother will not mention it unless asked if it existed; with this symptom in the adult is associated frontal headache; in the young child evidence of cerebral distress is shown by restlessness, peevishness, and drowsiness towards night, with a hot skin and some thirst. The motions are often ochrey and pasty in character, with a very offensive smell. The tongue is dry, red at the tip, and fissured. The urine scanty and highly colored. The pulse is variable, and it may rise or fall without affecting the heat of the skin, or without affording any prognostication that the child is better or worse. This is not so in typhus, where, if the pulse fall, the prognostic is good. It often happens that there is an increase of pyrexia towards night and remission in the morning. This is common, indeed, in all febrile affections in children; but in this disease it occurs with sufficient frequency to have given it the name of infantile remittent fever.

On the eighth or twelfth day of the disease a careful search will reveal the eruption. This consists of rose-colored spots, elevated slightly above the skin, disappearing on pressure. Each spot may last from two to five days, when it disappears, while fresh ones keep coming out. The spots occur on the abdomen, chest, and back; and their number altogether varies considerably, from two or three up to thirty or forty, nor does the number prognosticate anything of the gravity of the attack. Severer symptoms may now manifest themselves, such as vomiting, delirium, excessive diarrhœa; occasionally the opposite condition of obstinate constipation may exist. There is often some tenderness over the abdomen, particularly over the right iliac fossa, where gurgling may also be heard, or the abdomen may be tympanitic: sometimes the condition may be one of drowsy languor passing into heavy stupor, from which the child is with difficulty aroused. The child, during the progress of the disease, loses flesh fast, "wastes away." The face looks worn and anxious. Often there is cough, short and hacking, with some dyspnœa and harshness of breathing. Auscultation will reveal rhonchus and sibilus, and even large crepitation. Epistaxis occasionally occurs, and sometimes the gums bleed also. A crop of sudamina may break out; such an occurrence, however, will not *per se* affect the prognosis. The pulse often becomes reduplicate in character. The symptoms may either gradually improve in the third week, and the child become better, or the disease goes on, and may terminate in exhaustive hæmorrhage from the bowel, or perforation of the bowel, owing to the disease in the agminated patches of Peyer. If the pulse become reduplicate suddenly in the third week, especially with tenderness and great distention of the abdomen, hæmorrhage is to be feared. Hæmorrhage rarely occurs after the fourth week, but perforation may occur up to the sixth week. Muscular tremor, if out of proportion to the delirium, is a grave prognostic. Again, a very high pulse, say above 150, is a grave sign; in very young children this is, of course, not so. If headache and delirium coexist for some time it will probably indicate some cerebral mischief. Bronchitis and pneumonia may both arise, and should be looked for; they are both serious complications. If tuberculosis be present, typhoid often excites the active deposit of tubercle, and in this it differs from scarlatina markedly. Perforation may be preceded by some signs of peritonitis, with hiccough and vomiting, when a sudden paroxysm of intense abdominal pain indicates what has occurred. In such cases the perforation takes place in the small intestine, generally within a few inches of the ileo-cæcal valve. In children hæmorrhage is rarer than in adults. Dr. Murchison states that out of 232 cases under fifteen years of age observed by Messrs Taupin, Rilliet and Barthez, it occurred only once, but this *extreme* rarity is not borne out by the observations of other writers. Perforation is stated to occur in about 13 per cent. of the cases. It is certain that the solitary glands, which ulcerate later than the patches, are more liable to be attacked in children.

Post-mortem.—The agminated glands of Peyer are found in every stage of inflammation, from slight swelling or increased vascularity up to the severest form of ulceration and sloughing. The more destructive changes occur in those patches which are situated nearest to the ileo-cæcal valve. There is at first merely swelling of the mucous membrane over the patch; in the next stage the borders are elevated, the centre depressed, and small circular ulcers are visible, each corresponding to a closed follicle. If the ulcerative process extend through the peritoneal covering, perforation results. The solitary glands of the small intestines present similar changes. The mesenteric glands are also secondarily more or less congested, softened, and swollen. The spleen is commonly found enlarged, as it is often in all acute specific diseases, and its structure is softer and more friable than is natural. The liver is also found enlarged, and often softened. Besides these changes there may be congestion of the brain and its membranes, ulceration of the mucous membrane of the stomach, hepatization of the lungs; these are, however, the result of complications, and not pathological changes belonging to typhoid fever.

Diagnosis from common gastric disorder may be made by remembering that typhoid is rare in children under five years of age, rarer still in those under two, while slight gastric disorder incident to teething, &c., are at this age exceedingly common; further, the muscular weakness, the presence of fever (as shown by the thermometer), and the occurrence of delirium, will also aid the diagnosis. In tubercular peritonitis the tongue is generally clean and moist; there is no eruption, and the abdomen is distended, not from tympanites, but from serous effusion.

From acute tuberculosis it must be confessed that the diagnosis is often sufficiently difficult, still there will be the absence of eruption and usual absence of diarrhœa in tubercular disease, at any rate at the *commencement* of the illness, and, moreover, in it the stomach is generally flat and even shrunken; at the same time it must be remembered that typhoid fever undoubtedly disposes to the active deposition of tubercle, and hence cases may and do occur in which the diagnosis is extremely difficult or impossible during life. *Post-mortem.*—The tubercular ulcers of the agminated glands have a hard, thick, inflamed, elevated border, enclosing little yellow masses of tubercular matter attached to the base of the ulcer.

The diagnosis from typhus is given under the head of that disease.

The mortality is large, one-fifth of the cases attacked die, and it cannot be too often repeated that the prognosis must be guarded, as the fatal complications may occur just when convalescence seems about to take place. The most favorable prognostic is a fall in the temperature. The duration of the disease being from twenty-eight to thirty days, some guide to prognosis may be found in the time elapsed; for instance, great exhaustion about the twelfth or thirteenth day would be unfavorable, as there would be yet a long time to run, while a similar amount of exhaustion on the twenty-sixth or twenty-seventh day would not cause so much alarm.

Treatment.—At the commencement of an attack of typhoid fever it is of great importance to remember that from about twenty-one to twenty-eight days have to be got over, and that no treatment whatever can shorten the duration of the fever. And yet in no disease, perhaps, is the skill of the physician more needed or more shown. Of foremost importance will be to place the child in bed in a large well-ventilated room; the precautions against infection recommended under scarlatina should be practised; but it is well to bear in mind that typhoid is but slightly infectious as compared with that disease.

Then great cleanliness will be necessary, all stools and secretions must be immediately removed, and should be passed into vessels containing Condy's fluid or carbolic acid. The diet will require at first to be plain, simple, unstimulating, such as beef tea, veal tea, chicken broth; the farinacea must be charily given lest diarrhœa be provoked. Light puddings (without currants) and milk may be freely allowed. The use of wine requires the following cautions and restrictions: it will be wrong to give wine if the patient is doing well during the first three weeks of the disease, but if towards the twenty-eighth day there is much exhaustion wine will be required to get the patient safely through the disease. Its effects must be narrowly watched and the quantity nicely proportioned to the weakness of the child and the result produced. Nervous and muscular tremor are symptoms calling for the employment of wine, so again evident flagging of the heart's action is an indication of its requirement, but the rule is a sound one, that if there be any doubt as to the propriety of giving a stimulant it is better *not* to give it (the rule is reversed in typhus fever). As to medicine, purgatives are by all means to be avoided: a dose of calomel and jalap at the outset of typhoid fever may kill the patient; salines and all drastic cathartics are to be religiously avoided. If an aperient be from exceptional conditions imperatively called for, a small dose of castor-oil will be the best and safest medicament. It will be more often necessary and desirable to check the diarrhœa; for this purpose, mild doses of chalk and red gum, or sulphuric acid and opium, will suffice. Acetate of lead and opium is a good formula; or one-tenth grain of sulphate of copper may be given to a child five or six years old, or a starch-and-opium enema may be employed, three or four drops of Battley to an ounce of decoction of starch.

Then the abdomen may be covered with a moist warm flannel, well sprinkled with turpentine, or with a bran poultice. For symptoms of cerebral congestion and excitement the head must be shaved and cold cloths or ice applied. Bleeding in any form is inadmissible. Turpentine stupes or an enema containing turpentine will be the best remedies for tympanites. If hemorrhage occur, the case is not necessarily fatal, as such hemorrhage may be slight; and if promptly treated even when very severe it may be checked. For this a lead-and-opium injection, and internally gallic acid, or Tinct. Ferri Perchlor. may be useful, or an injection of Tinct. Ferri Perchlor. may be used ($\mathrm{m}$xv.— $\mathfrak{z}$ iv.), in an extreme case, and a bag of ice should be placed over the abdomen. The patient must keep absolutely still, even the urine should be drawn off—no movement, in fact, at all permitted. Should symptoms of perforation occur the bowels must be immediately locked up with starch-and-opium or lead-and-opium enemeta, and opiates in full doses by the mouth. In such and similar cases the form of opiate is often a question; the best are Battley, nepenthe, and Liq. Morphiæ Acetat. as liquids; or solid opium, as pill. When one is not borne another should be tried, and when it is desirable that the patient be kept under the influence of opium for some days, the form should be varied, as one loses its effect. It must be remembered that children do not bear narcotics well, and the effect produced must be narrowly watched. Mustard epithems or turpentine stupes may be called for if symptoms of intermittent pneumonia arise. Bronchitis will require an expectorant mixture of ipecacuanha, squill and senega, or if the phlegm be very tenacious and ropy, Ammon. Carb. will be useful. Lacto-phosphate of lime is highly recommended by M. Blacke alike as an aliment, and a medicament. It excites appetite and facilitates digestion. It is specially useful during convalescence.

Dr. Klein has recently announced the discovery of the actual fungus causing typhoid fever. It is a minute vegetable organism possessing mycelium threads of very unequal joints. I think that all such statements and discoveries should be received with extreme caution.

Dr. W. Strange of Worcester published an account of cases undoubtedly produced by what he calls "rat-broth," *i.e.*, water supplying the house coming through the bodies of decomposing rats which had died in the cistern. Where could Dr. Klein's mycelium and "micrococci" be in such cases? Take up any article on enteric fever published in the last six years, and find if you can any other that agrees with it theoretically. But practically the value of antiseptics, of sulpho-carbolate of soda, of hyposulphite of soda, and all such remedies is more and more brought home to us. It is not unlikely that "internal disinfecting," of which I have elsewhere spoken, will be largely adopted as greater experience shall confirm its value.

During convalescence the diet must be carefully attended to. The physician who remembers the tender condition of the lately ulcerated glands will not suffer his patients to imperil their lives by eating hard undigestible matters. Light puddings, rice, tapioca, custards, boiled fish, beef tea, light broths, such should be the diet of the typhoid convalescent. A *single error of diet* will not only bring back the diarrhœa, but may prove fatal. This cannot be too well considered and impressed upon friends and the patient himself, if old enough, as friends especially are very slow to believe it. By-and-by cod-liver oil, strong soups, and change of air will re-establish the strength.

Treatment by the cooled bath.—Perhaps no more convenient place may be found for alluding to the hydropathic treatment now largely in vogue for all febrile conditions in which hyperpyrexia occurs. One of the most recent papers on this subject, that by Dr. Binz of Bonn in the "Practitioner" for April, 1876, summarizes the matter thus:—If a patient at 104° Fah. is placed in a bath at a lower temperature he must part with heat. As the bath is cooled down by the addition of cold water, the temperature of the patient is found to have diminished. Cold baths, says Dr. Binz, have the greatest effect. They should be short and often repeated, weak patients should commence at 97° Fah. and the bath heat be lowered to 68° Fah. by the gradual addition of cold water. In the meantime the body *should be gently rubbed*—The italics are mine: I think the point important. Dr. Wilson Fox gave large quantities of brandy to his cases of acute rheumatism during their reduction of temperature, and in one case Dr. Fox expresses a doubt if the patient would have recovered without the brandy, and this case had twenty-four to twenty-eight ounces in the twenty-four hours. All this affords valuable indication for treatment. It is wise to remember that violent means, *however popular for a time, and even though supported by the best professional authority*, have never characterized what is best in scientific medicine. We need to be more and more aware of manias in medicine. Can the practitioner adopt a better motto than the Apostle's—" Prove all things, hold fast that which is good "? Certain I am that those who have witnessed the reaction—the rising again with redoubled vigor of the artificially lowered temperature—reminding one almost of the rush of pain and throbbing after the use of ether-spray, will prefer the "cooled bath," and that used cautiously, to the cold. Hæmorrhage from the bowels in typhoid fever, and perforation, are contra-indications against the use of "cooled baths" according to Dr. Binz and other authorities. Otherwise "every age and every constitution permits the withdrawal of fever heat." "For babies it is not necessary to go under 86° Fah. to have full effect." Cold affusion and cold wet-sheet packing are also recommended. I have said something about the latter when "medicated" in speaking of scarlatina.

5. Typhus Fever.

An acute, specific, contagious disease, lasting twenty-one days, characterized by a peculiar eruption, appearing between the fifth and eighth day, of which each spot is persistent.

This disease is stated to be more common in adults than in children. Such statement is probably erroneous; the disease, however, is generally milder in character in children.

Filth, overcrowding, bad ventilation, and all unhygienic conditions, favor the rapid spread of typhus, but none of these can generate the disease *de novo;* its essential propagator appears to be contagion. Like other acute specifics one attack usually confers immunity from future attacks.

The period of incubation is short—the exact time has not been ascertained; probably it seldom exceeds a week. Dr. Murchison, however, asserts that it is "about twelve days." The invasion is marked by headache, not necessarily frontal, with general uneasiness, feverishness, and, often in children, vomiting. Occasionally there are marked rigors. These symptoms increase in severity, and are accompanied by sleeplessness, thirst, high pulse, loaded tongue, and great prostration. The temperature rises at once, and attains a maximum of commonly from 104° to 105° F., though it may go a degree or two higher. The exacerbations are marked in the morning and still more so in the evening. About the seventh day remission often occurs, especially in favorable cases. The pulse in children commonly attains 140 or 150. A sudden fall in the pulse is a prognostic of death, or of some grave complications; in the latter case it rises again rapidly. The characteristic eruption which appears between the fifth and eighth day is often seen first on the back of the hand in the form of mulberry-colored maculæ, at first somewhat elevated, but in a day or two not so. In children the maculæ are fewer and less distinct, while a general mottling, irregular dusky red, and looking as if it were below the skin, and hence called subcuticular, is more general. The eruption in children often covers the whole body like measles; each macula is persistent, and fresh spots do not appear as in typhoid. The spots become ecchymotic, and the eruption may be out from two to three days to twelve or fourteen days, or even till the twenty-first day in the severest cases. There is no subsequent desquamation of the cuticle. As the disease advances, the mouth and tongue become dry, brown, and cracked, sordes form, and the breath has a distinct ammoniacal smell. Thirst is marked throughout. Diarrhœa may occur, but not so frequently, and not during the invasion period as in typhoid; oftener there is constipation. Bronchitis and pneumonia are common complications during the second week, and their symptoms should be watched for. Restlessness, sleeplessness, and delirium, are constant symptoms; the last specially marked in children. Convulsions occasionally occur, and are very fatal; they often accompany albuminous urine, and are followed by coma and death.

The sequelæ of typhus are few and of rare occurrence as compared with typhoid, or still more with scarlet fever. Lung consolidation, weak heart, swelling of the salivary glands, and very rarely erysipelas, are among the chief. Diagnosis from typhoid may be made by noting the points which are embraced in the following table:

Typhoid.	*Typhus.*
Diarrhœa—the rule.	Rare.
Stools—pultaceous, alkaline, and albuminous.	Less consistent, acid, and non-albuminous.
Intestinal hæmorrhage—common.	Rare.
Abdominal pain constant.	Rare.
Tympanites almost always present.	Rare.
Tongue—dry, cracked, thin.	Thick, dry, brown, not cracked, tremulous.
Epistaxis — comparatively common.	Very rare.
Eruption—well-defined margin, pink and papular, vanishing under pressure. Each spot lasts three days, and successive crops appear.	Less defined, irregular, mulberry colored; never papular; less elevated for the first day or two, disappearing on pressure, not so when the efflorescence has become hæmorrhagic. Each spot persistent—no crops.
Peritonitis — from perforated bowel.	Never.
Retention of urine—rare.	Sometimes.
Œdema glottidis—rare.	More common.
General convulsions—very rare.	Less rare.
Bronchitis occurs intercurrently very frequently.	Not so commonly.
Heart failure, uncommon.	Common.

The most remarkable post-mortem appearances are the changes in the cardiac tissues, which are soft and flabby, and often in a state of fatty degeneration. The blood is particularly liquid. There is also some effusion of serum in the ventricles of the brain. The spleen is often large, pulpy, and softened. The true maculæ, but not the mottling, are persistent after death.

Of children under ten the mortality is about 5 per cent.; between ten and twenty 8 per cent.; and the mortality increases with each decade.

Treatment.—Much will depend on the diet and judicious use of stimulants. The diet must be light and nourishing. As the appetite fails the skill of the nurse is manifested in judicious relays of beef tea, chicken broth, eggs beaten up with wine or milk, arrowroot and corn flour blanc manges, jellies, clear soups; food in some form is to be given little and often. For drinks, lemonade, tamarind, and black-currant water, orgeat, barley water with a little fruit essence in it, milk and soda water, are all useful.

The use of alcohol is one of the nice points of treatment; its judicious employment when required will save life, its employment when not needed will only do mischief. As a rule, children do *not* require wine; still the best rule for typhus is, when in doubt give wine, whereas in typhoid the reverse is the rule. Great prostration, rapid pulse, solidification of lung, are especial indications for stimulants. It is a bad practice to give wine *early* in the disease; the judicious physician will remember and count the days, and will hold his hand for a state of things at the seventh day which at the fourteenth he might meet readily by administration of wine; for wine in these continued fevers is a sheet-anchor, and if our sheet-anchor is over and the ship still drifts, there is no more hope; but if with our other anchors we bring her fairly under hand we have a resource for any sudden blast which may help us to ride out the storm. As to the form of stimulant for children, port wine or brandy are best, and the exact quantity should be noted as well as the time of exhibition. It is good practice in extreme cases to support the strength by enemata of wine and beef tea. For other treatment, cold to the head, careful regulation of the bowels according to their state, *e.g.*, enemas of one ounce of barley water with a few drops of laudanum for obstinate diarrhœa and tenesmus. By the mouth a combination of citrate and chlorate of potash; the latter in good doses is often useful. Carbolic acid is recommended by some writers. The sulpho-carbolate of sodium as an antiseptic deserves further trial. It is useful both for the patient and those around him. The weak mineral acids, sulphuric or hydrochloric, well sweetened and diluted, are serviceable; or hydrochlorate of ammonia, and if great prostration occur, and the heart evidently flags, carbonate of ammonia. In digitalis also a powerful heart stimulant exists, which rather controls and limits waste of vital action, and is specially applicable where rapid pulse and high temperature coexist. Bronchitis and pneumonia will call for a sinapism or turpentine stupe to the front or back of the chest as occasion may require. It is rarely necessary in children to empty the bladder by catheter, but the possibility of being so obliged should not be forgotten, and the urinary secretion should be watched. Citrate of potash in good doses is useful if the urine be diminished or very high colored. General hygienic conditions must be observed throughout as to the heating, lighting, and ventilating of the room—removing discharges, which should always be passed into vessels containing copperas or chloride of lime. Carbolic acid or Condy's fluid may be added to the water with which the body is sponged. Sulphur pastilles are useful fumigants. It is important not to attend fever patients on an empty stomach. For treatment by aconite and the "cooled bath," I must refer to what has been elsewhere said of these remedies.

6. Intermittent Fever—Ague.

This disease occurs but rarely in children, more rarely still in children under five, and as far as London is concerned a case is scarcely ever seen; a brief notice, therefore, will suffice. The disease in children assumes a very different aspect from its common condition in adults, for the paroxysms are not regular. Moreover, the immunity from suffering enjoyed by adults between the fits, in children scarcely exists; they are feverish, restless, and poorly all the time. Then, again, in children, the hot stage is greatly prolonged, and the sweating stage very imperfectly marked; and the rigors or cold stage may be altogether absent and replaced by great nervous depression, or even by convulsions. Hence the disease resembles genuine ague in very few points; perhaps scarcely enough for the identification of the disease. In children above seven or eight the true type usually manifests itself. Of the different forms of ague young children usually suffer from an irregular quotidian, and older children from tertian; the cause of the disease in children, as in adults, is the exposure to marsh miasma and to malaria. The disease is also commonest in spring and autumn. The cold stage is ushered in with coldness, shivering, the so-called "cutis anserina," or goose skin, blueness of the lips, diminished secretions, thirst, anxiety, hurried respiration, and small weak pulse. This may last from half an hour to four hours; from what has been said it will be understood that in children this stage is very imperfectly marked. The child is weak, low, restless, seems dull and heavy, or may have an actual and violent fit of convulsions. Then comes the hot stage, in which the skin becomes hot and dry; the temperature of the blood rises from $105°$, which is usually present during the whole of the cold stage (notwithstanding the feelings of the patient to the contrary), up to $107°$ or $108°$. Now he feels hot, the clothes are thrown off, the skin becomes red, swollen, and there is thirst, headache, and often vomiting; the pulse quick, full, and hard; respiration more regular. This stage in the adult lasts from two to eight or ten hours, and in the child is well marked and prolonged, with burning fever, suppressed secretions, flushed face and hot skin. Lastly in the adult comes the sweating stage, in which perspiration breaks out at first on the face and forehead, afterwards over the whole body. The temperature falls; the pulse becomes normal, and the respiration tranquil. The intervals, however, in the child are so feebly marked that it is often restless and poorly between the paroxysms. The splenic enlargement is well marked in children, and it is apt to be more permanent in character than in the adult.

Treatment.—Quinine fortunately exercises the same specific power in the ague of children which it exhibits in that of the adult. With its administration must be coupled a removal from the malarious spot to healthy air. Nor should the child be again suffered to reside in ague districts, as the tendency to recurrence is stronger in the child than in the adult. Warm clothing and generous diet will be needed after an attack of ague. Salicine, arsenic, and other antiperiodics are resorted to chiefly when quinine cannot be procured, or from some idiosyncrasy cannot be tolerated. Treatment is best commenced by a free purge, followed by the use of quinine in appropriate doses every three or four hours during the period of intermission. The body may be sponged with tepid water during the hot stage, and warm drinks allowed during the sweating stage. For the cold stage, warmth to the feet by a hot bottle, blankets, or hot air bath. The quinine will require to be given for some time after the attack to prevent recurrence. Bromide of potassium has been recommended for the so-called "ague cake" or enlargement of the spleen.

7. Variola (Smallpox).

An inoculable, contagious, and infectious disease, characterized by an acute febrile onset, followed by an eruption which is first papular, afterwards vesicular and pustular in the course of from eight to ten days.

Variola Discreta.—The period of incubation of this disease is twelve days, poorliness and malaise being its leading features; at its expiration come the symptoms of the eruptive fever. These are rigors, often very severe, vomiting and pain in the back, thirst and heat of skin. Pulse frequent; tongue furred; of these *pain in the back and vomiting* are exceedingly characteristic, in young children the pains in the back are not much complained of, and the judgment must be formed on the cerebral disturbance, which is generally considerable. There may be some sore throat at this stage, or delirium, or actual convulsions; hence the diagnosis may be still uncertain of the common mischief. At the end of forty-eight hours from the occurrence of the rigor (but occasionally both earlier and later) the eruption appears, first on the face and spreads downwards in another twenty-four hours. The febrile condition is *abated* on the breaking out of the eruption. The eruption itself appears first as small, red, slightly raised points, which enlarge and become papular; they feel hard like small shots; in three or four days a little lymph appears at the summit of each papule; by the fourth day, this continuing to enlarge, flattens at the top, and becomes umbilicated. The eruptive fever has by this time disappeared. There is an inflamed areola on which the vesicle stands. The lymph becomes purulent and as the pustule enlarges the umbilication disappears. The pustule attains its maturity by the eighth day; a dark spot may now be seen in the centre of each pustule, the inflamed areola subsides, and the pustules break and let out a liquid which forms a yellowish-brown crust and scab. If the pustules do not break they fall off as furfuraceous scales.

To return to the general symptoms that accompany the gradual maturation of the eruption; the face swells, the scalp becomes puffy. There is tension and burning in the face. A peculiar odor is emitted characteristic of the disease; there is much itching and tingling of the skin. The saliva becomes ropy, and there is often on the sixth day some swelling of the throat, hoarseness, and dysphagia. This indicates that the eruption has attacked the mucous membrane of the fauces, where, in fact, it may be seen in the form of round white spots. Similarly the eyelids, prepuce, and vulva become affected; this eruption is a day or two later in time than the general eruption and less pustular in character. About the eighth day, what is called secondary fever sets in, which is manifested by extreme jactitation, sleeplessness, quick pulse, scanty and high-colored urine, and delirium, especially at night.

In *Variola confluens* not only is the eruption confluent, as the name implies, but the primary and secondary fevers are both more intense. The primary fever does not abate during the appearance and maturation of the rash, and the secondary fever is severe and often typhoid in character. The pustules run into one another, so that large patches of pus may exist on the face or forehead. Various complications are also common in this form of the disease, as boils, abscesses, erysipelas, diffuse suppurations on the limbs and elsewhere, and blindness from affections of the conjunctivæ.

In the variety called *nigra* or *maligna*, the type of disease is lower still; there is more utter adynamia and prostration with delirium early passing into coma. The eruption often retrogrades or is associated with petechiæ. It is dark, even purple, in color. In this form hæmorrhages are common from the bowel, from the kidneys, and from the womb. In such cases the fever occasionally destroys life before the eruption appears.

The disease is contagious, inoculable, and infectious. It is important to remember that the patient is infectious until desquamation has taken place, and it is certain that clothes, &c., will retain the virus for years. It is a non-recurring disease. It is also important to remember that, as in all acute specific diseases, the severest form may be caught from the mildest.

Varioloid or modified smallpox is the disease occurring in those who have been vaccinated, or who have had a previous attack of general variola. In this disease the primary fever, though often very severe, does not last above one day, and may be followed by but a single pustule, or a few about the wrist and alæ of the nose; then the whole course of the disease is milder and less regular; the spots may appear in all their stages at the same time; the scars are slight; the odor is slight; and the secondary fever is generally wanting altogether.

Prognosis.—In variola discreta, perhaps one in four or five die. The disease is unfavorable in direct ratio to the amount of confluence. Children from nine to fifteen as a rule do well. Unfavorable signs are, the fever being typhoid in character, sudden retrogression of the eruption, many petechiæ, convulsions, and delirium, or complications with brain, throat, or lung affections. In infancy, the disease is dangerous, 50 per cent. under five die.

If death occur it is usually from the eighth to the thirteenth day.

Sequelæ are numerous and troublesome, besides the pitting and scarring of the skin, which varies with the quantity and nature of the eruption: these are often ulcers, boils, suppurating glands, erysipelas, deafness from suppuration of the internal ear, pleurisy running on to empyema, hæmoptysis, and hæmaturia, menorrhagia, &c. In the modified disease these never occur.

Post-mortem.—The skin presents the characteristic appearance already described. If the air passages have been affected, and death occurs on the eighth or ninth day, the mucous membrane will be congested and inflamed, and covered with a brown viscid mucous secretion; below this the mucous membrane is often ulcerated. If pleurisy or pneumonia have occurred their anatomical characteristics will be present. It is doubtful if the pustules of variola are ever seen in the gastro-intestinal mucous membrane.

Treatment.—The room in which the child is should be large and well ventilated, the temperature cool and diet low; and gentle saline aperients are to be administered. It may possibly happen that at the outset of the disease the cerebral congestion is so severe as to demand the application of leeches to the scalp. If this seem desirable a good number applied at once is better than a few repeated, and the bleeding should be stopped when the leeches fall off; in this way the quantity of blood taken can be more exactly determined, reckoning each leech to draw ℨij. Cerebral congestion is seldom present without hyperpyrexia. Present views would endorse the use of the "cooled bath" in preference to leeches. In fact, bleeding in all forms should be abandoned whenever possible; and if, on the other hand, the disease at once assumes the typhoid character, no time must be lost in the exhibition of stimulants and support. The hot bath will then prove useful to maintain the temperature and promote the throwing out of the eruption. In any case the course and progress of the disease will require to be closely watched, and the physician must be ready with his wine and nourishment the moment he perceives the powers of life to flag. As in all fevers, guidance to a successful course is to be accomplished by careful watching and prompt responses to the symptoms as they arise, rather than by attention to any preconceived rules and prescribed order of events. Cool drinks are, as a rule to be allowed, and tepid sponging of the surface. The hair is to be cut off.

It is recommended by Dr. Sansom and others to touch the apices of the pustules with pure carbolic acid, the odor of which may be disguised by a little oil of wild thyme. A solution of one part of carbolic acid in three of olive oil should then be applied over the individual pustules, night and morning. The general surface may be washed with coal-tar soap. Sulpho-carbolate or sulphite of sodium should be given internally. To complete the antiseptic treatment sulphurous acid vapor may be used to impregnate the air of the sick-room. I have not found this vapor to excite cough so much as would à priori be expected, and patients who cough a little soon become accustomed to the vapor even when pretty strong. The sulphurous acid spray may also be applied with benefit to the nares and pharynx. It removes disagreeable tastes and keeps the nose free from obstructions. Sulphurous acid may also be given internally, a teaspoonful in half a tumbler of iced water makes a good drink. The hyposulphite of soda has been used with success in the treatment of variola, ℨ ss. to ℨ j. doses being employed. The testimony of many good observers confirms the value of the "antiseptic" treatment of smallpox. Dr. Sansom gives the preference to the sulpho-carbolates, especially when head symptoms are present, and in "all zymotic ailments in which the throat is involved." The use of the sulpho-carbolates or other "antiseptic" treatment in no way precludes the use of aconite in half-drop or drop doses for hyperpyrexia, where the "cooled bath" is inapplicable or unobtainable.

For the sore throat, if the child be old enough, a mild gargle may be allowed, such as infusion of roses, or the mouth may be frequently washed out and cleansed by a syringe if necessary. If diarrhœa come on it must be checked. If abscesses form on the forehead or scalp they must be freely and early opened, while full diet, wine and quinine, will be called for. If the skin is very irritable and there is troublesome itching, sweet oil or sperm ointment may be gently rubbed in. It is a good plan to tie children's hands in a cloth to prevent their tearing and scratching, as they are sure to do. Sometimes a dry powder, as powdered starch or common flour, relieves the itching well. Pleurisy and pneumonia are very alarming complications; the former is almost always fatal. Blistering fluid should be painted on the side, and iodide of potassium given in good doses; wine and support should on no account be withheld. Pneumonia calls for sinapisms and the exhibition of Ammon. Sesquicarb. and Senega with Hydrarg. cum Cretâ night and morning, and rather less stimulating diet. Camphor is useful in combination with stimulants when there is subsultus tendinum, brownish tongue, and great debility combined with nervous prostration. For ophthalmia, no diminution in the amount of support is needed, a lotion of Zinc and Vin. Opii is useful, and the Ung. Hydrarg. Nitratis Mitius smeared between the lids at night. A weak lotion of nitrate of silver may be advantageously used if the conjunctiva get inflamed, with a small blister at the mastoid process, or on the temple. Strumous inflammation occurs especially in children, causing great photophobia; weak lotions of the Zinc and Vin. Opii, with steel and cod-liver oil internally, are the remedies.

To prevent Pitting.—It has been recommended to touch each pustule with nitrate of silver or camphor, or to bathe the face with a solution of four scruples of nitrate of silver to ℨ j. water. A mercurial plaster formed of—

Ung. Hydrarg.	25	parts.
Yellow wax.	10	"
Black pitch.	6	"

has a good effect. Dr. Aitken gives this as the formula used at the Children's Hospital in Paris.

Carron oil is a good application, till the scabs begin to loosen; they should always be removed when dry, or they stain the skin permanently Mr. Marson recommends cold cream and oxide of zinc, or if the discharge be thin and excoriating, calamine mixed with olive oil. Dr. W. Stokes uses light poultices over the face or a mask of lint soaked in glycerine and water and covered with a further mask of oiled silk.

8. Vaccinia—Cow-pox—Vaccination.

By Act of Parliament it is now ordered that every infant shall be vaccinated within three months of its birth, unless the state of its health should render the operation objectionable. On the introduction of the vaccine lymph into the arm of the infant no effect is noticed for a day or two, beyond the trifling blush occasioned by the punctures. At the end of the second day a small papule becomes perceptible; by the fifth or sixth this becomes vesicular and umbilicated; by the eighth day the vesicle is complete, no longer umbilicated, but full and round, of a clear pearl color. At this time, also, an areola or ring of inflammation surrounding the vesicle, and, spreading for the next two days, shows that the matter has affected the constitution, and is no longer a mere local disease. Slight constitutional symptoms are often manifested at the same time, as restlessness, slight feverishness, sometimes diarrhœa or sickness, or even swelling of the axillary glands. By the tenth day the areola fades and the vesicle dries; by the fourteenth day it becomes a mere scab, which contracts and gets darker, and ultimately falls off about the twentieth day, leaving a permanent depressed cicatrix of variable shape.

Vaccination properly performed, that is where four or five good vesicles have been produced and the areola is clear and satisfactory, affords protection up till the time of puberty, when it is desirable that revaccination should be performed. The phenomena of revaccination are similar to those of vaccination primarily, except that the vesicles are earlier by two or three days in attaining to maturity, a point of importance to remember. It will constantly happen that no second cow-pox can be caused, or a mere papule may form—so-called "Spurious vaccination," or occasionally the constitutional symptoms are severe. The more marked the "pits" in infancy, the less will be the probability of successful vaccination being performed a second time, the less will be the probability of the individual developing variola, and the more "modified," should he do so, will the variola be. Most remarkable were the statistics forthcoming from London, from Liverpool, from Paris, and other places during the recent (1870-71) epidemic. The rate of mortality from variola varies almost precisely with the quality and quantity of "good marks" on the arm. Hence it cannot be too much urged upon practitioners to vaccinate infants thoroughly, and not to submit them to the misfortune of an imperfect taking, which may be sufficient to hinder a subsequent taking, and yet insufficient to protect from variola. No good comes of vaccinating at every outbreak of smallpox. Such a course is rather to be deprecated, still less does good come of vaccinating after smallpox has developed itself. At the same time it is highly proper to revaccinate all who need it on an outbreak occurring in any house. Smallpox takes twelve days to incubate, and then forty-eight hours more of initiatory fever before the rash appears; revaccination up to the production of the areola, which is the criterion of its having become effectual, takes about seven or eight days, so that performed even four or five days after exposure to infection it may still prevent the disease or at least modify it. With those vaccinated for the first time two or three days more will be required for the areola to be perfectly formed, and therefore for protection to be secured. No one who consults the returns of the Smallpox Hospital, or the many other statistics available on this subject, can fail to be convinced of the efficiency of vaccination and revaccination when properly and *thoroughly* done. It is to the neglect of the careless, and the evasions which the sceptical practise in defiance of the law, to the imperfect performance of the operation, and to the silly clamor of ignorant persons, that we are indebted for the recent severe outbreak which, however, will not have been altogether without bearing valuable fruit, if it shall have served to silence foolish opposition by the eloquence of facts, and to stimulate indifference and carelessness by the presence of danger.

A child about to be vaccinated should be in good health, free from skin diseases, especially Lichen strophulus and herpes. If special reason exist, an infant may be vaccinated directly after birth; but about a month or six weeks old is the common and best time. The lymph about to be used should be taken from vesicles between the fifth and eighth day; perhaps the eighth is, on the whole, the best.

Cases are occasionally met with in which there seems to be an insusceptibility to the vaccine virus. I know a child in whom vaccination has been most carefully performed once a year for seven years, and she has never taken cow-pox; but such cases are rare, and failure is far more often from the operation not having been performed with sufficient skill.

In cases where a deficiency of vaccine lymph occurs, a condition of things by no means uncommon, particularly in the colonies, it is recommended to employ lymph diluted with glycerine. For example, one part of vaccine lymph to two parts of pure glycerine and two parts of distilled water carefully mixed in a watch-glass, is stated to be a most efficacious dilution. Dr. Wiener, of Cohn, vaccinated 1600 children with the above, and there were only five which "did not take."

9. VARICELLA—CHICKEN-POX,

Is a non-recurring contagious disease, accompanied with slight fever, and attended with a characteristic eruption of vesicles. Chicken-pox commonly occurs before the period of the first dentition. It is generally held that the disease is absolutely distinct and separate from smallpox, for it has been established beyond doubt that the occurrence of the one is not prophylactic against the occurrence of the other, and it is moreover incommunicable by inoculation. The prodromata of varicella, usually slight, may be smartly febrile in character, accompanied with drowsiness, and even coryza, so that the coming event cannot be prognosticated; but in twenty-four hours (sometimes as late as thirty-six or even forty-eight hours) the characteristic eruption appears in the form of little rose spots, acuminated, and from fifteen to twenty-eight in number, irregularly distributed, and even abundantly present, rarely confluent. On the second day a fresh crop appears, considerably more numerous, while the first spots have become filled with a clear serum. After twenty-four hours the contents of the vesicle become milky. The eruption is often attended with itching, so that the vesicle may be scratched open. By the fourth or fifth day it shrivels into a dry scab, and by the eighth or ninth the scab falls off, usually without leaving any scar. The diagnosis from variola and varioloid is made, not only by attention to the general symptoms, which are milder, but also the vesicles of varicella present no central depression as in smallpox, and the peculiar, hard, shot-like feeling of the variolous pustule is absent. Moreover, the variolous pustule is multilocular; the chicken-pox vesicle is unilocular, and collapses when pricked.

The disease is usually so mild, and is attended so rarely with serious complications or sequelæ, as to leave little to be said of treatment. Gentle saline aperients, avoidance of exposure to cold, and a warm bath towards the close of the affection, are the principal points to be attended to.

CHAPTER VI.

DISEASES OF THE BRAIN AND NERVOUS SYSTEM.

1. Idiocy and Mental Disorder.

An idiot is one who, in consequence of some cerebral abnormality originating before the brain has reached its full size, and the mind its full capacity, becomes irrecoverably deficient in mental power, and lacks the capacity to co-ordinate his brain functions.

Such brain abnormalities may be—
1. Arrest of development.   } Both of which may be secondary to
2. Arrest of growth.   }   some disease.
3. Disease, *e.g.*, chronic hæmorrhage into the meninges.

Mere backwardness must be distinguished from idiocy. This may be done by observing that there is no unusual size or shape of head, no fits, no paralysis, no spastic rigidity.

An excessive development of some normal attribute, *e.g.*, obstinacy of firmness, must not be mistaken for idiocy. Such cases require tact and great forbearance in their management. The child should not be curbed and threatened, but led into a better frame of mind.

So, again, mere idleness and inertness are not idiocy; but there is some danger lest they become such, because the brain is not duly exercised.

There is sometimes observed a temporary deficiency from nervous exhaustion and general debility, *e.g.*, on recovery after the acute specific diseases. This condition may last a long time, but it consoles the parents to know that it is always eventually recovered from.

Chorea is apt to degenerate into idiocy when very long continued; the stupidity which often exists during an ordinary attack of chorea is recovered from.

The question often arises, how shall an opinion be formed as to the state of the child's brain before it can talk? &c. Attention to the following points will generally solve the difficulty:

1. The child's eyes should follow a bright light or bright object in two weeks from birth; it should begin to smile about the same time. It should be remembered that squinting when objects are brought near them is natural and proper to children under one month old, but not afterwards; and because the child cannot at that age adjust its eyes it is no sign of cerebral disease.

2. A child should begin to use its hands and take hold at three months; to know familiar faces at three to four months; to know objects by name at eight to nine months.

The tongue should be kept within the mouth from the earliest age. The child should support its head at three months. Idiots always fail in this. The anterior fontanelle should close at from eighteen to twenty-four months.

3. A child should begin to talk at nine to sixteen months; to walk at ten to eighteen months; should feel its feet when held out to walk at nine months.

The child's brain should weigh at birth ¾ lb., at the end of five years 1·5 lbs.; by the seventh year it should have acquired its full size.

A small brain may be due to—
1. Deficient supply of blood.
2. Inflammation of the meninges.
3. Effusion of blood on the surface of the meninges in large quantities.

A large head may be due to—
1. Mere thickening of bone, as in idiocy and rickets. This is not a cause of idiocy, but secondary to it.
2. Pure hypertrophy. This does not cause idiocy, nor any symptoms until compression occurs.
3. Albuminosis of brain merely causes defective power, not idiocy.
4. Hydrocephalus often accompanied by idiocy.

Idiocy may be congenital or acquired; and whenever there is idiocy, whatever the cause, there is often some bodily defect, *e.g.*, arrest of development, fits, heart disease, spastic rigidity, and shortening of some muscles.

Hence the idiot cannot walk or talk properly; he is often deaf, cannot take hold of objects, and is also often deformed. His manners are childish or offensive; his expression is vacant: his ideas few. There is often much obstinacy, brutality, and dirty habits, the appetite greedy, the passions strong.

Treatment.—Much may be effected even in the education and training of idiots; in fact, it is surprising how wonderfully, under those who have the necessary patience, long-suffering, and experience, the poor idiot will develop into a being with some intelligence, and with trained and disciplined habits. This is not the place to enter into any details in the matter; it is sufficient to say that even congenital cases are not to be abandoned as hopeless and to refer those who would see and know more to the excellent establishment at Earlswood, Redhill.

2. CONVULSIONS.

M. Bouchut states that convulsions in children may occur in a state of health, and in the course of acute disease, and are then analogous to delirium; and there is no relation between such convulsions and lesions in the nervous centres; further, convulsions may, but more rarely, be symptomatic of genuine morbid conditions in the brain or spinal cord. At any rate, convulsions are of such frequent occurrence, and under such diverse conditions in children, that a careful inquiry into the various relations they present is most important for a due understanding of many of the commonest affections in children. The reason of their frequency in children as compared with adults is the predominance of the spinal over the cerebral system in early life. As the brain increases in size and power, convulsions become of rarer and rarer occurrence. Suppose a child actually in a fit, to which we are, as often happens, hastily summoned—a child that we have never seen before—what is to be done? and how is our diagnosis to be formed? In the first place it is well for one's own comfort, as well as for that of the friends, to remember that children do not usually die in convulsions, especially if we find the parents and friends alarmed, as they get used to the really dangerous forms, which are the ever-recurring ones; so we may reassure and comfort them. It is an excellent plan immediately to order a warm bath to be got ready, and to see also that the windows are open, or the door, so that the room is well ventilated, and to get the child removed from in front of a blazing fire, where it will often be found with its head within a few inches of the bars. Its nurse has thought that it felt so cold, so she proceeds to half roast it. Pass one hand quietly and carefully over the child's head, while the pulse is felt with the other. Fulness or weakness of the pulse will be a guide to diagnosis, so also if the head be really hot or cold; if the fontanelle be tense and protruding, or sunk and retracted; if the face be flushed or pale. Then order the child to be stripped, and observe if it draws its legs up to its belly; if so, and the head be hot and the fontanelle prominent, there is congestion, which may be the cause or the consequence of the convulsions. But in either case the immediate treatment will be to put its feet into a bath, into which may be thrown a handful of mustard, and at the same time apply vinegar and water or spirit and water lotion to the child's head. It is a good plan, also, to wring out a piece of flannel in the hot mustard-and-water bath, and to sprinkle a little more mustard on the surface of the flannel, and then wind it round each leg and foot.

If, on the contrary, the head be cold or the fontanelle depressed, it will be a good plan to pop the child altogether, except its head, into the mustard-and-water bath, and then employ friction to arouse the action of the skin. A little sal volatile may be held to the nose, and a few drops of brandy in a teaspoonful of water, to moisten the lips. Sometimes such a case is merely a syncope, and no convulsion has occurred. This must be remembered as possible, and looked for. It will be highly important during these proceedings to ascertain the previous health of the child. If this be the first fit it has had or not? Which of the acute specifics has it had? Is it teething? Pass the finger along the gums, and if swollen scarify them freely; and it is well to remember that convulsions sometimes occur in children with the *second* dentition as well as, very commonly, with the first. Another very important question is, What has the child had? If some unwholesome diet has been given, this is the cause of the fit, and a good purgative will cure it. Calomel and sugar is the best, as it can be put on the back of the tongue, and is sucked down without difficulty. Has the child vomited? If so, what has it brought up? and also the stools, if passed, should be looked at. The possibility of scarlatina, rubeola, or variola is not to be forgotten, in all of which the invasion may be by convulsions. Sometimes mere flatus is the cause of the fit; then the belly will be tumid, and gentle friction with the warm bath will dispel it. Having ascertained something of the previous history of the child, the knowledge may then be applied. If there has been diarrhœa, and the head is cool and the fontanelle depressed, there is no congestion and brandy may be given. If the child be emaciated, and the head hot, and diarrhœa have preceded, brandy may still be given, but cooling lotions may be required for the head. If the child has been irritable, and has had twitchings or "inward fits," the fontanelle being prominent, there is some abnormal condition of the nervous system. Calomel will be required at once, a few leeches, or perhaps a blister behind the ear or on the vertex. If there has been headache and vomiting and some feverishness, the pulse will be a most valuable guide; for while such symptoms, with a pulse of 130, would in all probability (as has been before mentioned) mean sweets, or plum cake, with a pulse at 40, they would be the earliest manifestation of tubercular meningitis. Should tubercular meningitis be the diagnosis, ice will be required to the head, calomel and jalap, &c. (See that disease.) Perhaps the child is recently convalescent of scarlatina, and has had some anasarca and a little albuminuria; if so, this fit of convulsions points to uræmic poisoning. The treatment will then be a hydragogue cathartic purge such as one-twelfth of a grain to one-sixth of a grain of elaterium, or twenty grains of jalap powder with a little scammony, to a child five years old, repeated every two or three hours. The loins should be dry-cupped, or a few leeches (to draw two or three ounces of blood) if thought desirable may be applied. Again, the paroxysm may have come on in the course of pertussis. There is then general congestion, and such congestions are often rapidly fatal. The treatment will be ice or cold to the head, free purgation, a counter-irritant, and the pertussis mixture should be a decided sedative to control the violence of the paroxysms. If these be very severe, hypodermic injection may be tried. If the physical signs of pneumonia be present, the case is also very serious. (See Pneumonia.) In measles, scarlatina, and variola, a convulsion at the outset is not a grave prodroma; in the course of the disease it is more serious, and it then represents the delirium of the adult, as before mentioned. Lastly, the convulsions may be caused by chronic hydrocephalus or hypertrophy (which see).

Death in cases of convulsions, when it occurs, may be from spasm of the glottis, asthenia, intense cerebral congestion and coma.

Dr. Gee, in a valuable paper published in the "Bartholomew's Hospital Reports" for the year 1867, gives statistics and details of 102 cases of convulsions; of these he refers 24 to local causes, *i.e.*, disease in or near the cerebrum, 73 to general causes, and of 61 children suffering from essential or eclamptic convulsions, he found 50 were rickety; and lastly, 5 cases he enters as of uncertain origin. Dr. Gee, in fact, seeks to prove the almost constant coincidence of spasm with the rickety diathesis.

A baby is often said to suffer from inward fits: this is when it lies as if asleep, but moves the eyelids, and the muscles of the face twitch slightly, and there is the so-called sardonic smile; the condition is generally due to flatulence and is easily relieved by gently rubbing the stomach and giving a few drops of tincture of cardamoms in some sweetened dill water. A worse prognosis must be formed, however, when the hands and feet are drawn in the so-called carpo-pedal twitchings, the eyes half closed, the child waking with a sudden start, the face flushed; then we have reason to fear an attack of general convulsions. The actual symptoms of such an attack are a terror-stricken look, twitchings of the face, rollings of the eyes, perhaps a squint or other deviation from the natural condition which gives a dreadful expression to the face, frothing at the mouth, the head and neck are drawn backwards or to one side, the muscles of the back are rigid, and the extremities violently thrown about. These movements may be limited to one side, and it is important to remember that such limitation *per se* does not imply organic lesion of the nervous centres. Consciousness and sensation are lost, the face becomes flushed, the eye insensible, and the pupils dilated or contracted, but in either case immovable; the breathing hurried and labored; the pulse quickened, small, hard, and often irregular. The urine and fæces are discharged unconsciously, and a clammy moisture breaks out over the whole body. This condition may last from a minute or two to an hour or more, when the child falls asleep or lies in a sort of stupor, or cries loudly, and returns slowly to consciousness, or sinks into coma. Except in pertussis, and laryngismus stridulus, and apoplexy, or after long exhausting diseases, a child rarely dies in a fit.

The above is a general sketch from which endless variations and deviations will be met with. There may be prodromata or none, the fit may be long or short, slight or severe, partial or general, recurring constantly or at rare intervals, all of which points will require careful notice in the individual case.

SPASTIC RIGIDITY (*Contraction with rigidity*).

This is an idiopathic muscular contraction of different flexor muscles of the extremities, especially of the fingers and toes, existing independently of any recognizable disease of the cerebro-spinal system. It often exists associated with laryngismus stridulus. It is most common between one and three years of age. It is commonly sympathetic in origin, though rarely it may be essential. The causes, when it is of sympathetic origin, are gastro-intestinal irritation, dentition, &c. The condition of spastic rigidity is also sometimes symptomatic of disease of the brain as tubercle and meningeal hemorrhage. When the disease is fully manifested, the thumbs are seen to be drawn into the palms of the hands, and the fingers, strongly flexed, cover and conceal the thumbs. This flexion of the fingers is at the metacarpo-phalangeal articulations, the phalanges themselves are extended and separate from each other. The contraction may extend to the wrists, forearm and even the arms. The toes likewise are in a state of muscular flexion or extension, and the foot extended upon the leg. It is rare for the spasm to extend to the knees. There is usually shooting pain and stiffness of the affected parts. There is generally much restlessness and irritability, but the mind is clear; convulsions, strabismus and other indications of nervous disorder may occur, but it is not very usual for them to do so. This condition may last for weeks or months either slowly increasing in severity or remaining *in statu quo*. Improvement is often manifested intermittently, the intermissions becoming longer and longer as restoration to health progresses. MM. Rilliet and Barthez mark the following distinctions between symptomatic and essential contraction:

Symptomatic contraction.	*Essential contraction.*
Cerebral symptoms and functional disturbances preceding or accompanying the contraction.	Cerebral symptoms not always present, and never preceding the contraction.
Frequent irregularity of pulse.	Never.
Usually partial, and commonly commencing in the elbows and knees and most often in a single extremity.	On both sides and commencing in fingers and toes.
Permanent as a rule.	Intermittent as a rule.

The prognosis is generally favorable. When death occurs, it is most frequently from convulsions.

The treatment will vary with the cause, which must be sought for, and first attended to, as for instance if it be gastro-intestinal or dental in origin. When the general condition has received attention, the local spasm will be best relieved by warm baths, the use of Bromide of Potassium in good doses or of Monobromide of Camphor, or of Belladonna, or Oxide of Zinc. No lowering measures are admissible, on the contrary the child's strength must be supported in every way.

3. Night Terrors.

This is a fruitful source of anxiety and distress to parents. I have certainly noticed that this condition seems to run in families, child after child of the same family becoming subject to it, whilst other children living in the same street or square, or even in the same house, brought up under conditions almost precisely similar, remain completely exempt. The child goes to bed quite well to all appearances, but in two or three hours after it has been asleep, it suddenly awakes in great alarm and gives utterance to loud and terror-stricken cries. For a few minutes it may fail to recognize its nurse or mother; it will point to the bottom of the bed or under the clothes with an expression of great alarm, and will often imagine some object hanging near, as, for instance, a part of a dress or shawl, to be some animal about to attack it. Presently it gets more composed, bursts into tears, and sobs itself to sleep in its mother's arms, and the attack may not return for some nights, or it may recur with wonderful periodicity night after night about the same hour. I have known children thus affected weeks at a time, when intervals of quietness would take place, perhaps to be again broken into after the lapse of a month or two. Fortunately, this condition, alarming and distressing as it is, does not usually, if ever, depend on cerebral disease, but seems to be entirely of gastric origin. Sometimes pale urine, like the urine of hysterical women, is voided freely after an attack. Often dentition is present, and sometimes there is great constipation.

The treatment of these cases will comprise, in the first place, kindness and forbearance towards the little sufferer. These terrors will be but increased by harshness, while soothing and gentleness will do much to dispel them. The child should by no means sleep alone, its cot should be beside the bed of its nurse or mother. A light should be left in the room, and when the fit occurs it should be soothed and encouraged. The gums, if swollen and tender, should be freely lanced, but it is mere cruelty to lance gums which do not require it. A combination of tonics and aperients will be found valuable, as tending to correct the condition of gastric and intestinal disturbance which almost always coexists. The diet should be carefully regulated, simple, nourishing and easy of digestion. Iodide of Potassium is often useful when combined with an aperient; and I have also seen great good from the compound syrup of the Phosphate of Iron.

4. Congestion of the Brain.

This condition is of such frequent occurrence in early infancy and childhood, and its consequences are so important, that it behoves the practitioner to watch carefully for its symptoms, and to be ever ready to meet them when they arise. There are two kinds of congestion: 1. Active congestion, which is that in which the vessels of the brain become overloaded from an increased flow of blood to them. 2. Passive congestion, in which some mechanical or other impediment exists to the return flow of blood from the brain.

1. *Active congestion.*—This condition may result from any of the fol-

lowing causes: The onset of one of the eruptive fevers. The irritation of dentition. Exposure to the direct rays of the sun, or, what is more common, the foolish plan that many mothers adopt of sitting to wash and dress their children almost in the fire, with the head of the unfortunate child as near the bars as a joint is in roasting, and this to avoid risk of cold.

Gastric disorder is another source of active congestion.

Symptoms.—The onset may be sudden or gradual, general uneasiness, restlessness, sleeplessness, with some amount of fever, and generally constipation, may exist for a few days; or suddenly, perhaps, the child awakes from sleep with a scream, his head is noticed to be hot, the face flushed, he vomits. If old enough he complains of his head; if younger, the anterior fontanelle is seen to be prominent, and on feeling it, it pulsates strongly. The pulse is quick, the muscles of the face work and twitch. There is jactitation, often some delirium, or in a few hours an attack of convulsions passing into stupor. This condition of congestion at the onset of the eruptive fevers, say scarlatina, for example, may be so severe as to kill the child within twenty-four hours and before the appearance of any rash, and it may only be discovered that such was the fatal cause by the simultaneous outbreak of scarlatina, after similar but less violent prodromata in another child of the same family. Or the attack may be less severe, and in a day or two out comes the rash, or the tooth comes through, or, in other words, the cause passes away and the head symptoms cease. Here, then, is a condition alarming, rapidly tending to get worse, but relieved almost as rapidly when the exciting cause has passed away. Such cases tax the judgment and coolness of the physician, who must determine upon his treatment at once, and act calmly and decisively when all around are excited and anxious, proposing all sorts of remedies and expedients, and questioning him again and again as to his opinion respecting the termination of the attack. The first thing to be done is to ascertain if possible the cause, and the finger should be passed along the gums; if swollen and tender they should be lanced freely. It is well to cause a warm bath to be got ready, as during its preparation, time is gained to ask a few questions of the circumstances immediately preceding the attack. Has the child been out? If so, was the sun hot? What has it had to eat? and when? And this question must be pressed on the nurse and servants as well as on the mother. While out, nurses constantly give children sweets and trash of which these attacks are often the revealers. It is well, therefore, to assure the nurse that the child's life depends on the correctness of the information given to guide us, and she will generally disclose the truth. It would be dreadful to be leeching and blistering the head for fancied hydrocephalus or inflammation, when an emetic and a dose of calomel and jalap would clear away the mischief in the form of plum cake. If no tangible cause can be ascertained, the eruptive fevers should be borne in mind—which of these has the child had? Has it been exposed to infection? If *no* clue can be got, and the congestion be *very* violent, and the child strong and healthy, six leeches should be applied over the scalp. It is always wise, in cerebral affection more particularly, to determine the quantity of blood to be drawn, to reckon each leech at ℨij., and then when the leeches fall off, to stop the bleeding. A purge of calomel and jalap will be useful in unloading the intestinal canal. A blister behind the ears or on the vertex is sometimes useful (it should be made with the blistering fluid). Ice to the head and cold affusion are also valuable measures, and may often replace leeches and all such treatment with advantage, while Bromide of Potassium may be given internally, with or without Aconite. It is of the greatest importance, whatever line of treatment be adopted, to darken the room, to leave the windows and door open to allow cold fresh air freely to enter, and to enforce absolute and complete silence and quietness. In cases of milder nature and in delicate children, a gentle purgative, a warm bath, and mustard pediluvium will be advantageous. When the immediate danger is passed, treatment will generally be required for a few days to complete the cure and prevent a relapse. Salines with an occasional gray powder are suitable remedies.

5. Passive Congestion.

This occurs in the progress of such diseases as pertussis and laryngismus stridulus, in hypertrophy of the liver and spleen, from an enlarged thymus, and in children dying from asthenia.

Symptoms.—A soft but weak pulse, puffiness, and blueness of the face, moist cool skin, cold hands and feet, uneasiness, or pain in the head. With these symptoms there are often conjoined clayey stools, very offensive in character, perhaps some diarrhœa, with occasional vomiting and general loss of appetite. During a paroxysm of pertussis such a child not unfrequently dies of coma. In such a case the vessels of the brain and its membranes are found loaded with black blood quite fluid, the choroid plexuses are highly congested, and on a section of the brain being made more bloody points than are natural appear.

Treatment.—If the cause be pertussis or laryngismus stridulus, no treatment will be of use which does not cure or relieve them, which should be the first consideration. Depletion in any form, even a few leeches, are very badly borne in such cases as these; temporary good may result, to be followed only by worse weakness and exhaustion. In a severe case counter-irritation is stated to be of value, such as a small blister behind the ears. More often the best treatment will be the daily use of a warm bath, containing a little mustard and some Tidman's sea-salt, while cold is applied to the head. Alterative aperients alone will often work wonders. Gray powder, which is given too often "usque ad nauseam" for the diseases of children, is in this disease a valuable remedy, a small powder containing Hyd. cum Cretâ and a few grains of rhubarb, and P. Jalap. Co., if necessary, should be given every night, and some of the Syr. Ferri Iodidi three times a day. This plan will be beneficial, even when clayey diarrhœa coexists, as it restores healthy action of the intestinal canal and at the same time gives tone to the system. The diet must be carefully regulated. As a rule it will be found that beef tea, mutton and veal broth, all increase the diarrhœa, while arrowroot and corn-flour are not properly digested. The best foods will be milk, rice-milk, raw meat, shredded fine, in small quantities, light puddings, and no wine. If stimulant is required it may be given in the form of medicine or brandy. The child should be warmly clad, and should have plenty of fresh air.

6. Hæmorrhage.

Cerebral apoplexy, or effusion of blood into the substance of the brain, is in children a rare disease. The symptoms are similar to those observed in the adult—heaviness, drowsiness, and headache, passing slowly into stupor, or the attack may be sudden and appear as coma, convulsions, or paralysis. *Post-mortem.*—A clot may be found in the brain, most commonly about the corpora striata and optic thalami. There is sometimes a certain amount of softening around the clot, as a consequence of the effusion. *Meningeal apoplexy* is the more common affection, and, according to Cruveilhier, constitutes one-third of the cases of death in stillborn children. The effusion is into the cavity of the arachnoid; it is very rare to find hæmorrhage into the ventricles. This form of apoplexy is seldom accompanied by paralysis.

The onset is often in an attack of convulsions, which frequently return, and there is also spasmodic contraction of the feet and hands, strabismus, vomiting, great thirst, fever, and duskiness of the face. The convulsions continue to recur with increasing frequency, and soon close the scene. Sometimes the attack may be more sudden—drowsiness, followed at once by stupor, coma, and convulsions. Two forms depend on the relative amounts of effusion; when this has been great, death takes place very rapidly. It is well to bear in mind that passive hæmorrhage may take place into the cavity of the arachnoid from long-continued and exhausting illness; in such cases the symptoms are very insidious. *Post-mortem.*—The blood effused undergoes certain changes, according to the time elapsed since its effusion. At first it is fluid, by the fourth or fifth day coagulated, the serum is absorbed, and the clot becomes adherent to the parietal serous membrane. The color fades, and the clot at first becomes a thin, fibrinous lamella, looking like a false membrane. It is generally impossible to detect any opening through which the blood escaped. In children with tubercle of the brain it is not uncommon to find limited apoplexy, consisting of innumerable bloody points. This has been called "capillary apoplexy." The *prognosis* of the disease, in any form, is most unfavorable. The *diagnosis* of meningeal apoplexy from acute meningitis is mainly to be made by observing that the symptoms are less inflammatory, the invasion more sudden, and the loss of voluntary power more complete. Chronic hydrocephalus is slower of development, and is accompanied by prodromata before effusion, while in meningeal apoplexy effusion symptoms are the first observed. Paralysis will aid in diagnosing the cerebral form.

Treatment.—This, of course, will vary with the cause and kind of apoplexy; if the symptoms be those of congestion, a few leeches over the scalp and perhaps blisters behind the ears, followed by calomel and enemata; but if, as more often happens, effusion has already taken place, such measures will be hurtful, and we must trust to enemata, ice to the head, and sinapisms to the feet. If the condition of the child be that of great exhaustion, with feeble pulse, and clammy skin, a warm bath and mustard poultice to the chest, gentle friction to the body, and the careful use of stimulants, afford the only chance. Alarming symptoms are often relieved at the period of dentition by free scarification of the gums, with a brisk purge and an occasional use of the "cooled bath."

7. Tubercular Meningitis (Acute Hydrocephalus).

This dire disease most frequently attacks children under five years old. The symptoms are, for purpose of convenience, divisible into three stages.

The *first* or precursory stage is ushered in by some or other of the following group of symptoms:—Irritability and capriciousness; headache and fulness, shown by the child frequently putting its hand to its head, or by the head hanging down, sleepiness and drowsiness, occasional dragging of one leg, disordered appetite, vomiting, constipation, elevated temperature, and disturbed sleep; the stools pale, clayey and offensive; the tongue moist, red at tip and edges, and furred in the centre; the pulse quickened, seldom, however, above 120, and often irregular; photophobia. The child during sleep does not close its eyes, grinds its teeth, and often wakes in alarm, is paler than natural, though there may be a transient flush, a peculiar irritative cough is a common and significant percursor. A pinched, drawn, haggard expression is very characteristic. It occurs even early in the precursory stage.

Even at this stage the respiration is quickened a little, is unequal and irregular, and accompanied with sighing and yawning; the temperature is usually increased, though not so much as in other tubercular diseases.

The recognition of this stage of the disease, the duration of which is usually from four to five days, is of the utmost importance; during it only has treatment any reasonable hope of success.[*]

The supervention of the *second* stage is marked by increased moroseness of the child: it wishes to be left alone; at night there is often considerable delirium; the pulse is slower, falls, perhaps to 80 or even 40, is more irregular, and even intermitting; slight exertions, however, materially quicken it for a time; there is more stupor and insensibility; the child frowns almost constantly; the face is flushed; there is much heat of head and pulsation of the anterior fontanelle; there is heard the peculiar piercing cry called the "cri hydrocéphalique." The pupils are often unequally dilated, or there may be strabismus. The abdomen is remarkably shrunken. This stage passes by insensible gradations into the *third*, which is marked by increase in the stupor, often broken into, however, by convulsions. The convulsions may leave paralysis, usually the same, sometimes of the opposite, side. The pulse becomes small, rapid, scarcely to be numbered. There are clammy sweats. The pupils are widely dilated and motionless. The aspect of the little sufferer is piteous to behold, with shrunken face and form, eyes staring and sunk deep in their sockets. Convulsions constantly recur, and soon put an end to the scene. It happens, however, sometimes that an improvement takes place for some days before death. This is a significant fact in the course of many chronic as well as acute diseases. Just before death a remission of, perhaps, the very worst symptoms takes place. The pain, which has been agonizing, vanishes; the breathing, which had been so labored, gets easier; the purging, which had been so uncontrollable, ceases; and the patient's friends delude themselves with false hopes. The physician must ever bear this in mind; it is remarkably conspicuous in many diseases of children, and in none more, perhaps, than in tubercular meningitis. Ourselves undeceived by it, we must caution the friends from expecting permanent amendment; such second disappointments sicken the heart.

Exceptional conditions.—The symptoms of tubercular meningitis are exceedingly variable, both in character and sequence; and while the above may be considered a correct sketch of the true type, almost every case will vary from it in some particular.

---

[*] To M. Bouchut belongs the credit of first employing the ophthalmoscope in the detection of the earliest stages of this disease. The test is not infallible, as optic neuritis does not invariably occur. When, however, the following appearances can be detected, they are characteristic: 1. Congestion around the papilla, with patches of congestion on the retina and choroid. 2. Changes in the retinal veins around the papilla, as dilatation, tortuousness, varicosity, thrombosis, and patches of hæmorrhage from rupture.

The most constant and persistent symptoms are vomiting, constipation and retraction of the abdomen. Hemiplegia, ptosis, and strabismus are not uncommon. Trousseau considers as diagnostic of tubercular meningitis, the so-called "tache meningitique," a red line remaining on the skin after the finger had been drawn along it. This sign, however, is not peculiar to tubercular meningitis, it occurs in other conditions of cerebral congestion and even in pneumonia. But it does not occur in typhoid fever, which is sometimes a point of value in diagnosis. Some cases have been recorded in which convulsions set in from the very first, or others in which they occurred only at the last. In some, while the pain in the head is the earliest and most conspicuous feature, in others it may be absent altogether; and so with vomiting, the dilated pupils, and strabismus; and yet there is a general resemblance in the course of the different classes of cases which experience recognizes at once, and the symptoms never fail to point to the brain as the organ at fault. The onset of tubercular meningitis is often most insidious, and will require careful attention to the premonitory symptoms to detect its probable approach.

Diagnosis.—From typhoid fever. In the latter disease the following points contrast with the ordinary course of tubercular meningitis. It is common in children above five years old. There is often no vomiting. The bowels are relaxed. There is tenderness and gurgling, especially in the right iliac fossa, with a tumid abdomen, and abundance of flatus. The tongue is dry. There is more heat of skin, and no irregularity of the pulse. Convulsions and paralysis are rare. From simple acute meningitis M. Rilliet draws the following distinctions; it must, however, be admitted that in most cases the two affections are almost indistinguishable:

1. Tubercular meningitis occurs in weak, precocious children, and in those subject to glandular enlargements and skin diseases, whereas in simple meningitis the objects are vigorous and healthy.

2. Tubercular meningitis is always sporadic.

3. The child previously pines away, and suffers from gastro-intestinal irritation. Simple meningitis begins without prodromata.

4. Tubercular meningitis does not commence with convulsions.

5. In simple meningitis headache is if possible more intense, vomiting more urgent, constipation less obstinate, fever more violent, delirium higher.

6. In tubercular meningitis the progress is comparatively slower.

7. The duration is more prolonged.

8. In simple meningitis the disease is more ataxic from the first, and the aggravations more progressive and continuous.

Prognosis extremely unfavorable; few cases recover in which the first stage is past. The duration of the disease is from ten to twenty days.

Post-mortem.—Traces of inflammation of the membranes of the brain are found with the results of such, viz., serum, lymph, and pus. The dura mater may be healthy or injected. The arachnoid is injected and often opaque, dry, and sticky; there is also effusion of a clear fluid between the pia mater and arachnoid, and more rarely pus. At the base of the brain there is also opacity of the pia mater and arachnoid, and effusion of fluid between them and in the meshes of the pia mater, but in addition in the membranes, the arachnoid especially, are found minute tubercular granular deposits, yellowish and friable, or gray, opaque, and firm. The central portions of the brain are softened and often contain tubercular deposits. These granulations are found to be identical in microscopical appearance and chemical composition with ordinary tubercle. Lastly, there is besides the softening (which may be even to a mere creamy consistence) of the central portions of the brain, an effusion of watery serum into the lateral ventricles; this fluid often amounts to several ounces. The lining membrane of the ventricles is also thickened and opaque, and its vessels full and turgid; occasionally it presents a granular appearance. Tubercular deposits are also generally found in other organs of the body, especially in the lungs and bronchial glands, less often in the liver, spleen and mesenteric glands, and softening of the stomach occurs in about two-thirds of the cases.

It is convenient here to mention the form of hydrocephalus called by Gölis "water-stroke," which consists in sudden effusion of fluid into the brain, which occurs either idiopathically, or as a result of obstructed secretion from some other organ, or as a secondary affection in the course of one of the acute specific diseases. In such cases death is too rapid for the employment of remedies.

Treatment.—1. Prophylactic. When one child of a family has died of hydrocephalus the health of the father and mother should be inquired into and improved as far as may be practicable. The mother should in future not be permitted to suckle, but the infant must be reared by a good wet-nurse. Besides this, every hygienic condition should be brought to bear upon such an infant. His food, his clothing, his exercise, must all be carefully considered and adapted to his growing necessities. Sea air will always be beneficial, and baths in salt and water as soon as the child is old enough. The diet must be plain, simple and nourishing; stimulants are undesirable. Such a child must be allowed to be backward in his lessons; all attempts at forcing his intellect must be discouraged, his health must be the first and only consideration. The parents must be made to understand the importance of things apparently trifling in his case, such as a little vomiting or constipation. Such a child should never have "home medicines" administered to him. The Cetraria Islandica moss is a valuable adjunct in diet, and the Syr. Ferri Iodid., or compound syrup of the phosphate of iron with cod-liver oil the best medicines. The bowels must be most carefully regulated, and every tendency to any disorder narrowly watched; this will be done by a wise mother without allowing her child to perceive that it is an object of undue solicitude, with quiet undemonstrative attention, without any fussy, foolish interference. Great tact, great forbearance, and firmness will be needed in the education and management of such a child. 2. If the disease be actually established the treatment becomes an anxious and much disputed question. I do not believe that bleeding in any form can be beneficial in the vast majority of cases. The object of depletion is merely to relieve congestion, and this may be effected far more satisfactorily by applying cold to the head, an evaporating lotion, wet rags, ice—all are beneficial. The hair should be cut off.

In a case of simple acute meningitis it is possible that leeches may be called for; they are, indeed, recommended, and have been used by Rilliet and other good physicians. In the tubercular form I am quite sure that bleeding and lowering measures are very badly borne; counter-irritation is the means by which we must seek to give relief; this may be accomplished in many ways. A couple of blisters behind the ears applied *early*, before the effusion stage, sinapisms to the feet and legs, and small flying blisters, are useful and better than one or two large ones, large blistered surfaces being often very slow to heal. Calomel is another remedy on which we are usually taught to rely. I am quite satisfied that in many cases calomel and all forms of mercury do positive harm. The tubercular diathesis does not indeed so strongly contra-indicate the use of mercury as the scrofulous, but I prefer to administer such purgatives as secure a copious watery secretion from the intestines—one of the best is compound jalap powder, it may be combined with a few grains of P. Scammon. Co., this, however, is not to be pushed too far, but the constipation naturally existing is often obstinate and will require careful and well-regulated doses of the aperients to overcome it; emetics are in my opinion quite inadmissible. Any one who has taken an emetic knows the fulness of the head which retching occasions, and will perceive that such a condition cannot fail to increase the cerebral congestion already existing, besides vomiting and nausea are amongst the most prominent and troublesome natural symptoms, and no good can be done by still further exciting them. If much febrile disturbance coexist I am in the habit of combining with the compound jalap powder, Pulv. Jacobi ver. in moderate doses. I am quite sure that it exercises a beneficial effect, but the medicine in which I am disposed to place the greatest reliance, and which I have seen to be often exceedingly beneficial, is iodide of potassium. I believe that the reason that many practitioners have failed to obtain benefit from its use is because they have been giving calomel and gray powder in various doses two or three times a day, which does as much and more harm than the iodide of potassium can do good.

There remain one or two remedies which have been employed and require brief notice. Digitalis, for example, of which I have no experience, has been found useful by some physicians. Opium, the value of which is much disputed, and the chief use of which appears to be to tranquillize in cases of unusual cerebral excitement. To sum up the general course of treatment it seems, on the whole, that the best results are obtained from moderate counter-irritation, active purging, and the internal administration of iodide of potassium in doses of two or three grains thrice daily (for a child three or four years old). Bromide of potassium or bromide of ammonium will replace the iodide with benefit when convulsions occur.

Elaterium and croton oil have been recommended in the obstinate constipation of this disease. I can only say that their use will require the greatest caution, and, of the two, I think croton oil the less objectionable, and the more certain in its action. In a disease where treatment of all kinds is so very unsatisfactory, I would advocate a further trial of the tinctures of gelsemin and scutellarin.

I have made pretty extensive trial of the concentrated tinctures of B. Keith and Co., of New York, and I am persuaded that in the "Eclectic Medicines" we have remedies too little known, and the value of which will be more and more appreciated. I have found the extremely disagreeable taste of some of the concentrated tinctures a disadvantage, but it can generally be surmounted by glycerine, oil of lemon, or other devices of elegant pharmacy. In children this is essential. We have no right to give children filthy doses when a little skill and consideration can make this nauseous drug pleasant. With this "caveat" I earnestly advocate a tentative use of some of these potent agents. Gelsemin is a nervine tonic of marked value in many convulsive affections. Scutellarin soothes and quiets the irritability of the nervous system, lessens cerebral excitement, and at the same time excites diaphoresis and diuresis. Their alternative use, and their use in combination I am testing as opportunity offers. The dose of Keith's tincture of gelsemin for a child five or six years old is one drop, the same of scutellarin. Both drugs can be given as powders if preferred.

The diet is to be nourishing, certainly not too low; stimulants are, I think, uncalled for in the earlier stages of the disease at any rate. The food is best given in small quantities and often. Free purgation often checks the vomiting, which otherwise is a serious hindrance to the administration of food. Ice may be sucked when the vomiting is very obstinate, or sinapisms over the stomach may be tried. In extremely obstinate vomiting the hypodermic injection of small quantities of morphia over the stomach is, in my opinion, very valuable. The room in which the child is should be darkened, but well ventilated. Quietness is to be strictly enjoined; emphatically, such a child must be prohibited all excitement, soothing and judicious nursing being of the utmost importance.

9. Hydrocephaloid Disease.

This is a disease which it is simply of vital importance not to confound with true hydrocephalus. It is essentially a disease not of inflammation, but of debility, from loss of blood and other causes. Dr. Marshall Hall, who was the first to point out the nature of this condition, divides it into two stages. First, that of irritability; second, that of stupor. In the first stage the infant becomes irritable, restless, peevish; the face flushed, the surface hot, the pulse frequent; there is an undue sensitiveness of the nerves of feeling, so that the child starts and cries at a noise or on being touched; there are sighing, moaning, and even screaming during sleep. The bowels are flatulent and loose, and the stools offensive. If, through any erroneous diagnosis, stimulants and support are withheld, or if diarrhœa supervene, the exhaustion developed leads into a yet worse train of symptoms. The face becomes pale and cool, the eyelids half closed, the eyes wandering, the pupils insensible to light, the breathing becomes irregular and sighing, the voice husky, often with short cough. The stools are green; the feet cold. These symptoms especially supervene in newly weaned infants, fed with improper food, and also in the enfeebled condition left after exhausting treatment, and also in infantile diarrhœa. The practical matter is to ascertain in all cases of doubtful head symptoms in young infants the previous history; whether any other children of the family have had hydrocephalus or are tubercular. Whether the child has been lately losing flesh, if it has been lately weaned, and, if so, on what it is fed. Also, if diarrhœa has existed, and if it vomits, and when—that is, if only after it takes food—or at other times when the stomach is empty. A common source of error is admitted by all writers, and that is when real congestion has been over-leeched and blistered, and this form of disease sets in as the result. This even may be guarded against by observing that in spurious hydrocephalus the face is pale and cool, and the anterior fontanelle sunk instead of protruding and pulsating.

Treatment.—When this condition is discovered, the child must at once be put on good diet. Exhausting treatment, if such were being employed, immediately suspended and stimulants supplied. If the stomach be very irritable, small quantities of ass's milk with a little lime water in it will be found the most useful food for an infant, and veal tea or chicken broth without fat for an older one. If the child be very low, a mustard bath may serve to rouse the system, and enable it to swallow a little brandy and other stimulants.

Opiates are of value in quieting the nervous irritability as well as the diarrhœa that often coexists. It is hardly necessary to say that opium must be given with caution. The state of the pupils affords, according to Dr. Churchilll, the best guide. When it is dilated, opium enough to overcome the dilatation and make the child sleep should be given; smaller doses merely increase the excitement. Dr. Churchill also mentions a very severe case in which, after the failure of anodynes, he had recourse to the inhalation of ether (thirty drops in a handkerchief) with marked success, quiet sleep supervening, and the child doing well after. Tonics are often serviceable, steel and quinine are best.

10. Simple Encephalitis

is a very rare disease in children except as the result of direct injury. It is less frequent in infancy than in childhood. The disease may set in suddenly with an attack of convulsions, which may be partial or general, and followed by some loss in the power of articulation—the face and eyes being distorted, and considerable stupor supervening. This is broken into by a fresh attack of convulsions, followed perhaps by complete hemiplegia, strabismus, and insensibility. In a few days death supervenes. Sometimes the course is not so rapid, and somewhat different; the disease may set in with disordered health, loss of appetite, deranged bowels, with frequent vomiting, some confusion and stupor, the eyes are heavy, the pupils dilated; a convulsion may not take place for some days, to be followed, however, on its appearance by fatal coma. Sir Thomas Watson considers the nausea and vomiting at the commencement point to the cerebral mass as the part affected; while convulsions indicate meningeal mischief.

Post-mortem.—There is found congestion of the vessels of the brain and general vascularity of its substance. The pia mater is especially vascular; there is also serous effusion into the ventricles, and beneath the pia mater; sometimes flakes of lymph are contained in the serum and the membranes may be coated with lymph and pus, the membranes are thickened, and there is softening of the cerebral substance. The sinuses are full of coagulated blood.

Treatment.—First comes the question of early depletion; to be satisfactory in its result it is essential that the bleeding should be performed *at the very onset* of the disease, and is generally most suitable in cases occurring from direct injury; if delayed it will but serve to lower the patient, in whom irreparable mischief has already taken place. Seen, then, at the commencement of the attack, a dozen leeches may be at once applied to the scalp, and calomel and jalap given both as a derivative and purgative. In my own practice, as is elsewhere often stated in this book, leeches and blisters are extremely rarely employed. The diet must be moderate and unstimulating, cautiously improving as the cerebral mischief appears diminished. Ice and cold to the head are measures very valuable in the less severe forms of the disease which are far more common, and in other cases good adjuncts after depletion. The exhibition of small doses of calomel or Hyd. c. Cretâ frequently repeated, may be useful in the course of the disease.

Cold affusion, that is, water steadily and continuously poured over the head, is a very powerful antiphlogistic and is stated to be effectual in this disease; the effect must always be watched, as it is an extremely depressing remedy, used for a few minutes it often suffices in congestion to afford marked relief. Iodide of potassium is a drug from the administration of which in fair doses some good may be effected, and aconite in quarter- or half-drop doses every half hour, is a very serviceable remedy in this as in other acute inflammations. Considerable lowering of the pulse and diaphoresis are indications for withdrawing or lessening the aconite. I have often known it necessary to do so after a very few doses.

Convalescence in such cases will require the greatest care and watching to prevent a relapse, all sources of excitement must be carefully avoided. Bromide of potassium is often a useful agent in procuring sleep, and calming undue excitement following an attack of inflammation of the brain.

11. Hypertrophy of the Brain.

This is usually met with in the rickety and scrofulous, and in infants six or eight months old. It is important that this disease should be recognized from hydrocephalus, to which it bears some similarity. This may be done by careful attention to the following points. The symptoms of chronic hydrocephalus come on earlier, and earlier grow more serious, especially in cerebral disturbances. Again, in hypertrophy neither is the head so large as in hydrocephalus, nor are the fontanelles and sutures so widely open; and, moreover, the enlargement in hypertrophy is mainly occipital. The downward expression of the eye, so characteristic of hydrocephalus, is wanting in hypertrophy; and instead of being tense and prominent, as in hydrocephalus, the anterior fontanelle and sutures rather sink and show depression in hypertrophy. Again, in hypertrophy the child lies horizontally, or throws back the head, with which, especially with the occiput, it bores into the pillow. In hydrocephalus the posture is commonly prone, and the head lower than the rest of the body and buried in the pillow. Dyspnœa is a prominent symptom in hypertrophy, not so in hydrocephalus, in which cerebral disturbance is the earliest sign of functional derangement. In hypertrophy the child is often fat, while in hydrocephalus it is thin and aged-looking. In hypertrophy the child's head is almost constantly in a state of profuse perspiration.

There is in hypertrophy a sensation of firmness communicated to the finger on pressure being made over the fontanelles. Children, the subjects of hypertrophy, are usually dull and drowsy, apathetic, irritable. Greedy appetite, vertigo, and frequent headaches are also concomitant symptoms.

Post-mortem.—The brain is found free from fluid in the ventricles, the gray matter is but little altered, and the general appearance of the brain is pale and anæmic (unless some cerebral congestive attack has been the immediate cause of death); the white matter is pale and firmer than natural and increased in size, according to Rokitansky, by an albuminoid infiltration into the granular matter between the nerve-fibres. These changes are noticed in the hemispheres especially, and do not affect either the cerebellum or the base of the brain. The thymus gland is often enlarged, and the dyspnœa has been referred by some writers to thymic asthma.

The disease is not necessarily fatal, though it may be so, especially if the sutures are ossified, from compression, and a fit of spasmodic convulsions may terminate life; or the child may recover and grow up an idiot, which is happily rare. Dr. West points out that cretinism and hypertrophy of the brain are often associated. The child is frequently carried off by some intercurrent disease, *e.g.*, pertussis, scarlatina, &c., in most of which cases the cerebral congestion will be found to have been unduly increased

MM. Rilliet and Barthez mention an extraordinary effect of lead preparations in producing this disease.

Treatment will be directed chiefly to maintaining and improving the general health; remembering the pallid condition of the brain bleeding is out of the question. Again, these children being either rickety or scrofulous are bad subjects for calomel. Counter-irritation, as in Dr. Churchill's plan of painting the head with tincture of iodine, will probably be of benefit. The hair should be shaved before the paint is applied; others recommend issues and setons, with more questionable propriety. Iodide of potassium should be given in fair doses. Dr. Elsässer employs a small horsehair cushion for the child's head to rest upon, a hole being cut for the occiput; this contrivance is said to cause the boring movements of the occiput to cease, and to afford the child some quiet sleep. Remembering the profuse perspiration of the head, it should be covered at night with a thin cap to prevent risk of cold; by day the scalp may be sponged with cold water. As the disease seems checked more tonic treatment may be ventured on, such as quinine and steel. The child's diet should be moderate, and such as it can readily digest. Free purgation should be directed against any symptoms of cerebral congestion that may arise.

12. Chronic Hydrocephalus

may be congenital or acquired; in either case the indications of cerebral mischief are plainly perceptible; the head may or may not be found enlarged, and if not enlarged at birth it soon becomes so. Convulsions constantly recur. The child presents a pitiable aspect; it loses flesh daily, although it may suck heartily; the cry is harsh and hoarse; the eyes roll frequently; there is strabismus; the legs are doubled up on the belly and the feet obstinately crossed. Moreover the feet and hands are cold. The fontanelles and sutures are widely open. The cause of death is often an attack of convulsions.

In the acquired form, at first the symptoms may not be very striking; in fact, it sometimes happens that the various functions are but little altered and impaired, and that even the mental powers are not greatly weakened until a short time before death; while in other cases the constitutional symptoms are urgent, and complete idiocy exists.

Perhaps the first noticeable signs are loss of muscular power, drowsiness, and some dulling of the intellectual powers; the child does not notice as he was wont, and he "always seems for sleep." By-and-by this is changed to moroseness, evident headache, photophobia, great fretfulness, emaciation, the limbs twitch and there may be actual convulsions in infants; in older children these are rarer, but there is in them a constant and peculiar frown, grinding of the teeth at night, obstinate constipation and scanty urine, with dilated pupils, squinting, or the sight may be altogether lost; the state of the mind varies from complete idiocy up to a condition in which it seems but little affected. The most remarkable feature of the disease is the cranial enlargement. In infants this is rapid, owing to the ready separation of the sutures, but enlargement takes place even when the sutures are ossified. It is the vault of the cranium that enlarges, the base remains always unaltered. The face does not in any way participate in the enlargement; the bones of the skull, viz., the frontal, parietal, and superior part of the occipital, with a small part of the squamous portion of the temporal, become expanded and thinned; sometimes they are like pieces of parchment, and the fluctuation of the water within can be distinctly felt. The fontanelle is elevated in rickety enlargement, it is depressed by the changes of the angle of the superior with the orbital portion of the frontal bone, the eye is driven down and half concealed by the lower lid; this gives a peculiar and characteristic appearance to the face.

As the head enlarges so its weight becomes more insupportable, and hence it rolls from side to side, and hangs down on the child's shoulders for support. There is generally loss of all power of locomotion, and often paralysis. The organs of respiration, circulation, and digestion are the last to be affected, for, though occasionally vomiting may take place, the child is generally voracious, but after a time these two fail; the breathing becomes labored, there is distaste for food, and the child may die after a year or two, or, in rare cases, a sort of living death may be prolonged into many years. If the attack have come on after ossification of the sutures, death is more rapid, because the brain is more compressed. Hydrocephalic children are also often carried off by some intercurrent disease. The irritation of teething is again a very fatal time with them.

Post-mortem.—The fluid may be found, according to M. Breschet, between the dura mater and the cranium; between the dura mater and the arachnoid; in the cavity of the arachnoid; in the meshes of the pia mater; and, lastly, and most commonly, in the ventricles. The site of the fluid will, of course, materially affect the other morbid conditions, compression from without would reduce the brain to a small size, and it has been found the size of a large nut; or, again, distention from within will render it a thin membranous bag, as it has also been found. It is often difficult to recognize the difference between the white and gray portions. The membranes are found either quite free from change, or showing signs of inflammation, especially that lining the ventricles has been found thickened and covered with granulations. The disease is attributed by some writers, as, for example, by Dr. Battersby, to an arrest of development. Dr. West, on the other hand, regards it as the result of slow inflammation of the arachnoid, and especially of that lining the ventricles.

Treatment.—All plans are sufficiently unsatisfactory. The plan of Gölis is perhaps the one that has been most frequently adopted. It consists in the administration of calomel in quarter- to half-grain doses twice a day, together with the inunction of one or two drachms of mercurial ointment into the shaven scalp once in the day; at the same time a flannel cap is to be worn to prevent chill. If no improvement take place at the end of six or eight weeks, diuretics are to be given, and an issue inserted in the neck or shoulders.

Of course it is necessary to remember that some cases, especially congenital ones, and cases depending on malformation, are not susceptible of cure.

Another plan is that of bandaging the head with flannel bands or strips of diachylon plaster; releasing it of course if symptoms of compression arise, which they sometimes do.

A third plan is that of puncturing and evacuating the fluid. This is performed at the coronal suture with a small trocar and canula about an inch and a half from the anterior fontanelle, care being taken to avoid the longitudinal sinus—a portion of the fluid only is removed at a time, and pressure carefully maintained between the tappings. It is, however, evident that this method is not applicable to the ventricular form of the disease, but only to those of the so-called external hydrocephalus. Legendre gives a sign by which external hydrocephalus can be diagnosed—" that it is never congenital, but begins usually about the tenth month, or about the period of the first dentition. The head enlarges gradually and does not attain the size it does in the ventricular disease, and it is, moreover, preceded by repeated convulsions and other forms of cerebral disturbance." There are also other points against the operation, as, for example, whether the fluid will not collect again, as in other dropsies · whether even if it did not, the state of the brain is such as to allow of reasonable hope of its recovering its functions. There is also danger of wounding some vessel of the brain, and, moreover, inflammation may set up as a consequence of its performance.

Dr. Churchill gives a long list of unsuccessful cases, and also points out how frequently unsuccessful cases are not recorded. Still the fact remains that the operation has been occasionally successful; perhaps one case in fourteen has been thus cured. At any rate it should not be tried till other methods have failed, and then only in carefully selected cases.

After tapping, injections of iodine, strong and weak, have been employed as it would appear without exciting inflammatory mischief. Thus, Dr. Tournesko of Bucharest injected ʒ iij. of tincture of iodine in ʒ v. of distilled water after tapping and drawing off 24 ounces of serum. After 15 days the child seemed in excellent health, the girth of the head being diminished from 56 to 43 centimètres.

My own practice consists in the employment of blistering fluid, applied sometimes behind the ears, sometimes on the scalp. With this I administer the iodide of potassium in large doses, and very frequently I combine it with the Syr. Ferri Phosphat. Co. Several patients of mine at the Victoria Hospital improved remarkably under this treatment. Iodide of iron is also efficacious. The expediency of employing mercury is much disputed. No doubt its use is occasionally attended with excellent results, but probably, in the majority of cases, it is better avoided. Sir Thomas Watson records two cases in which ten grains of crude mercury were rubbed down with ℈j. of manna and five grains of fresh squills, and given every eight hours, and this treatment proved successful in both instances when other means had failed. It was continued for three or four weeks and caused copious diuresis, emaciation, and debility, but no ptyalism. When the hydrocephalus disappeared, the strength was restored under the use of Griffiths' mixture. Diuretics are recommended, the concentrated tincture of eupurpurin (Eupatorium purpureum) in four-drop doses will be found effectual for this purpose. The drug is stated to have some special value in chronic hydrocephalus; of this I have no experience, but, as mentioned under tubercular meningitis, I am fully alive to the value of scutellarin, and gelsemin in many disorders of the nervous system. In children, scutellarin is particularly useful in controlling restlessness, wakefulness, and nervous excitability, as well as actual convulsions. These are valuable properties in many of the cerebral disorders. Where one child has been born hydrocephalic it is recommended to give phosphate of lime to the mother, throughout any subsequent utero gestation. It is certain that the general health of the father and mother should be well inquired into, and if any remediable cachexia exist it should be cured.

13. Tubercle in the Brain.

Very rare in adults; is not a very rare affection in children. MM. Rilliet and Barthez found tubercle in the brains of 37 out of 312 children between the ages of 1 and 15, in some organ or other of whose body tubercle was also deposited. The tubercular masses vary in size from a millet seed to a large walnut; in number, from a single mass to 40 or 50; the tubercle is sometimes crude; sometimes softened in the centre; generally globular; often very firm and less friable than in the lungs or lymphatic glands. The brain substance around crude tubercle is usually unchanged. It is common for the tumors to be attached to the pia mater; but they may be found quite unconnected with the membranes. The hemispheres, the cerebellum, and the pons varolii, are all frequent sites of their occurrence. These tubercular deposits are often associated with evidences of meningitis, as, for example, a granular condition of the membranes, an effusion of hyaline matter into the pia mater at the base of the skull, and the ventricles are also often distended with serum. It has been suggested that mechanical obstructions offered to the circulation by the pressure of the tubercular masses, is the explanation of this. It is rare for tubercle to be confined to the brain, it almost always coexists in other viscera, all the depositions being equally evidences of the diathesis, *i.e.*, the tubercular.

Symptoms.—Unfortunately, these are very obscure, sometimes there are none at all, and tubercle in the brain is only discovered after death. At other times, headache is the only symptom; it may be expressed in the young child by drowsiness and listlessness; it is not uncommon for the headache to be of acute, intense, lancinating character, at times, or at others to be completely absent; such attacks combined with vomiting are exceedingly suspicious. Deafness, and discharges from the ears are very common. Amaurosis is not rare. Convulsions occur especially when the tubercles are in the central portion of the cerebral mass. Rigidity, or contraction of one or other extremity, is not uncommon. Dr. West points out that convulsions may occur on one side only and yet the tubercle be found on both sides of the brain; or the convulsions may be general, and the tubercle only on one side. In a word, no disease presents symptoms more variable, because of the many secondary affections excited, *e.g.*, encephalitis, tubercular meningitis, &c., and also because of the variety produced by the different position and condition of the tubercular matter, which may be crude, softening, or cretaceous. The following summary will afford the best clue to diagnosis.

1. The age: the disease being commonest before three years.
2. The general history, if tubercular or otherwise.
3. Headache, irregularity of the pulse, and coincident otorrhœa.
4. Tubercular infiltration of other organs.
5. The chronic course of the disease as contrasted with the acute course of other cerebral affections in childhood.

The prognosis is grave, but not wholly hopeless, remembering the cretaceous state in which the tubercle is sometimes found after death.

Treatment.—Improving the general health and treating symptoms as they arise, are the two chief indications, for instance, gastro-intestinal irritation must be checked by gentle aperients, regulated diet, &c.; and attacks of cerebral congestion call for purgation, cold to the head, and moderate counter-irritation. These cases are, in fact, more susceptible of improvement by treatment than might be supposed, but it would be impossible in a disease of such protean manifestations to lay down exact rules. Cod-liver oil and steel are of course valuable. Iodide of iron with an occasional aperient, moderate counter-irritation, and well-regulated diet, is the proper treatment of many cases. Bromide of potassium is valuable when there are convulsions or general irritability. Phosphorus is a remedy valuable in this, as in many other atonic and tubercular affections.

The syrup of the hypophosphite of lime, or the fiftieth of a grain of pure phosphorus, made into a pill with three grains of reduced iron—one such pill may be given to a child from seven to ten years old, twice a day after meals. Cocoa, chocolate, and Iceland moss are highly nutritious substances, of much value in the diet of these difficult and trying cases, in which a change often seems so beneficial to the child when sickening of its ordinary food.

14. Chorea, or St. Vitus's Dance.

This disease is not absolutely limited to childhood, but it is commonest between the ages of seven and fifteen years. The disease may either set in suddenly, or with some gastro-intestinal disorder and irritability of temper. This is followed by slight twitching, convulsive movements of the face or one of the lower extremities; these gradually increase in severity, extending to the other limbs and even to the tongue. This condition of course disturbs all the voluntary movements; such a child cannot walk without a jerking movement and staggering; he cannot protrude his tongue without great effort, when it is suddenly shot out; he cannot stand still, but fidgets perpetually; he cannot take hold properly of anything, he misses his aim, and makes his efforts with the most strange grimaces.

As a rule (to which, however, there are many exceptions) the movements stop in sleep. In long persistent cases the intellect is apt to be seriously affected, and it is generally dulled during the progress of even an ordinary case. The general health is often but slightly impaired, but there is almost always *constipation*, and sometimes loss of appetite, and other gastro-intestinal derangements. The disease is commonest in girls, and is often observed about the period of puberty, when it may be connected with retarded or disordered menstruation.

The specific gravity of the urine is notably increased at the height of an attack of chorea, and declines as the disease declines. There is a real and very remarkable connection between chorea and rheumatism. Children who have had rheumatic fever are eminently prone to be attacked by chorea. The exact proportion of the cases so complicated has not yet been fully determined, nor is any assignable cause known for the connection. M. Sée states that of 109 cases of rheumatism admitted into the Hôpital des Enfans, 61 were complicated with chorea. Other writers have not noted so large a proportion. The late Dr. Hillier states that of 37 cases, he found 15 who had themselves been rheumatic, and 7 others, one of whose parents was said to be rheumatic.

My own experience would lead me to believe that M. Sée's figures indicate by no means too great a proportion. At the same time, in some districts and countries the relationship seems not to exist; for instance M. Rilliet says that rheumatism is very common at Geneva, and chorea exceedingly rare. M. Lombard says that in a practice of twenty years he saw but one case of such complication. Dr. West says that, whereas his opinion was formerly against such a complication, the returns of the Registrar of the Children's Hospital, in Ormond Street, in which, of 33 children affected with chorea, 11 presented a rheumatic history, his views have changed in the matter, and he admits the relationship is a very real one. But more remarkable, pathologically, than rheumatism, is the cardiac complication that so frequently attends chorea; this is oftenest manifested by a systolic murmur at the heart's apex, sometimes associated with considerable irregularity of the rhythm, and greatly disordered and tumultuous action of the heart. This murmur is described as dynamic, and attributed by Dr. Walshe, not to inflammation, nor to any organic change of the mitral valve. "Neither have such murmurs the usual accompaniments of hæmic murmurs, but they do seem plausibly attributable to the disordered action of the muscular apparatus connected with the valve." With regard to the frequency of cardiac complication in chorea, the late Dr. Hillier states that of 37 cases in his note-book, there was probably organic disease of the heart in 25; in 4 others, evidence of functional derangement, and in 8 only no signs of cardiac disturbance. Dr. Hillier considered that dynamic apex murmurs in chorea are rare. He rather attributed the choreic murmurs to organic change, because in his experience their subsequent disappearance had been rare. I have, however, had recently under my care two remarkable cases in which most distinct mitral murmurs completely passed away on recovery from the chorea.

Choreic bruit is often connected with vegetations on the valves of the heart. Sometimes it is evidently due to pure anæmia, and sometimes, especially where palpitation exists, to the irregular contraction of the walls of the heart. Rilliet and Barthez state that an intercurrent attack of one of the acute specific diseases diminishes or cures the chorea; out of 19 cases 9 were attacked by intercurrent diseases, and 8 were affected thereby; in some the chorea diminished from the first, in others it diminished more slowly or even appeared at first increased, but it afterwards disappeared.

Dr. West observes that he has known cases in which a cardiac murmur so arising during the course of chorea became persistent, and remained as a permanent condition, and, moreover, was followed by obvious dilatation of the heart, proving rapidly fatal. The usual termination of idiopathic chorea is in recovery, but it may prove fatal either by exhaustion or coma. It occasionally gives rise to some organic change in the brain, leading to convulsions, palsy, hydrocephalus, and apoplexy; these are, however, very rare. Relapses are apt to take place, and the disease frequently recurs even several times. It is not uncommon for chorea to be complicated with hysteria. Chorea is occasionally the result of embolism of a cerebral vessel, according to Dr. Hughlings Jackson of the small vessels supplying the corpus striatum, and neighboring convolutions.

Post-mortem.—The appearances found are rather those of the secondary affection, which causes death, than of the chorea itself; in fact, death has occurred in cases of chorea without any discoverable organic lesion. It is usual for the vessels of the cord to be congested, and there is sometimes an effusion of blood around the theca.

The membranes of the brain also are frequently affected, and there may be evidence of exudation, or serous effusion; occasionally softening of the brain and cord has been found. Dr. Russell Reynolds regards chorea as a brain disease rather than an affection of the spinal cord, because (1) Tonic spasm, rather than choreic clonic spasm, is the result of spinal irritation; (2) The choreic movements are in some degree under the power of the will; (3) The movements cease in sleep, whilst excito-motor actions are increased by the withdrawal of the will; (4) On the other hand, efforts of will specially increase the choreic movements.

Causes.—The disease is commonest in girls; three girls suffer to one boy. Children of nervous and hysterical women are more prone to suffer from it. The disease has been stated to occur epidemically, but the influence of irritation must be remembered as in the case of hysteria. The exciting causes are a blow, fright, worms, dentition, mental excitement, and any irritation of the nervous system. The late Dr. Anstie considered insufficient food a main cause of chorea, and proper and continued nutrition its cure in such cases. He calls special attention to the gravity of cases, occurring in young girls after the catamenia have appeared in excess, or in some other abnormal manner. The duration of the disease is from two to three weeks to as many months. Jaccoud observes that the period of three months marks the limit at which chorea passes into the intractable type. It has been supposed that some cases of stuttering, and persistent winking in adult life are in reality choreic, that is "local chorea."

Treatment.—The various plans of treatment consist in the administration of tonics, aperients, and anti-spasmodics. Remembering how common is constipation in this disease, the value of aperients is not to be lost sight of. At the same time an abuse of purgatives in a disease essentially cachectic is to be carefully guarded against. Simple remedies to regulate, rather than to purge, are best. Steel has been generally recommended as one of the best tonics. In my hands the Syr. Ferri Phosphat. Co. has often proved most valuable, subduing the worst features of the disease in about two weeks. I have also found the bromides of potassium and ammonium very useful. M. Gaillard considers that the bromide checks the violent and inordinate movements, and that it is useful in such rather than in the ordinary cases of chorea. Either drug should be given in fairly large doses, *e.g.*, gr. v.—x., for a child ten years old. In severe cases the child should be kept in bed. Probably the best plan is to commence with a gentle aperient, and repeat it two or three times a week, to keep up a regular action of the bowels. On the day after the first powder commence with 3 ss. of the Syr. Ferri Phosphat. Co., giving it thrice daily, and increasing the dose up to 3 j. If in two or three weeks no amendment takes place, or if for some reason steel is inadvisable, scutellarin, bromide of potassium, especially when dysmenorrhœa is present, belladonna, Liq. Arsenicalis, Succus Conii, or strychnia may be tried. Succus Conii was given to a child six years old, by Dr. Anstie to the extent of *eight ounces* daily, without producing any more effect than if so much water had been given. I have seen a similar result from extravagant doses of henbane, and conium in the adult, and we know the half-ounce dose of digitalis in delirium tremens. Such results would not deter me from employing these drugs in ordinary doses, in cases in which I had satisfied myself of their value, and I am further persuaded that small, rather than large doses of many vegetable sedatives are effective. Dr. West gives Liquor Strychniæ (B. P.) ℳv. 6tis horis to a child seven years old, and gradually increases the dose to ℳx. Other remedies much extolled are valerian (especially the ammoniated tincture), oxide of zinc, antimony in cases attended with febrile excitement, and nitrate of silver. Sir Thomas Watson combines cannabis Indica with small doses of nitrate of silver, and a little dilute nitric acid. Opium and stramonium have been used, and cimicifugin is recommended by American writers. It may be given either as saturated tincture, of which the dose is a drachm thrice daily, or as decoction (made from half an ounce of root to one pint of boiling water), of which from four ounces to half a pint should be drunk daily.

Trousseau uses sulphate of strychnia one grain in three and a half ounces of syrup; of this two drachms three times a day, to be increased by small quantities until itching of the scalp and slight muscular stiffness are observed, or six drops of the ordinary tincture of Nux vomica may be given three times a day. Arsenic is often very useful; Dr. Radcliffe recommends the hypodermic use of Fowler's solution. Its ordinary use by the mouth I have found to be serviceable in cases in which other remedies had failed.

Dr. Copland recommends a cold shower-bath, the child the while standing in warm water. The late M. Baudelocque used warm sulphur baths—such are made by adding sulphuret of potassium to the ordinary bath heated to 90° F. The child should remain in half an hour daily. M. Jaccoud's treatment consists in using the ether spray apparatus, over the whole length of the spine, twice a day. This is akin to the spinal ice-bag treatment of Dr. Chapman. Both appear to be well worthy of further trial. Gymnastic exercises, counter-irritants, and electricity have also been recommended. The continuous current has been found the best form of electrical action, the positive over the affected muscles, the negative conductor to the nerve supplying them. In a disease which occasionally proves extremely obstinate it is well to have a wide range of weapons with which to combat it. When obviously dependent on worms, dentition, or amenorrhœa, it is hardly necessary to say that these points must receive most careful attention, and be cured before we can hope for the abatement of the chorea.

15. Eclampsia Nutans, or Salaam Convulsion.

This disease occasionally affects the mind permanently. Its distinguishing character is the peculiar bowing forward and downward of the head so as nearly to touch the knees; this movement, at first slow, increases in speed, and may be repeated 50 to 100 times in succession. The attack often occurs on awakening from sleep. Sooner or later other convulsive movements occur, and the attack may end in paralysis or idiocy. Mr. Newnham inclines to the belief that this disease is of a strumous inflammatory character, the membranes investing the medulla oblongata being especially affected. The treatment will consist in careful attention to the general health, in the administration of sedatives, especially the bromide of potassium, and also of tonics, especially zinc and quinine.

16. Epilepsy.

In a good proportion of cases hereditary tendency to this disease may be traced. The exciting causes are fright, injuries, gastro-intestinal disturbances, dentition, &c.; dentition is probably the commonest cause. It should be remembered that the mind in children, being undeveloped, is more likely to be permanently affected than in adults. In the female the occurrence before menstruation more usually tends to cause insanity than its occurrence after that function has been established. The disease is more common in females after the age of seven years than in males. The disease, so far as the symptoms in the actual fit are concerned, do not materially differ from those observed in the adult. There may or may not be premonitory warnings; of course the child is often too young to interpret them even when they are present—the convulsive movements, the bitten tongue, and the subsequent coma, the flushed face, and the fixed pupil, and the labored breathing, all occur in the child as in the adult. The duration of the fit may be from five to eight minutes, up to half an hour or more. In children the character of the seizure is that of the 'petit mal.'

Post-mortem.—In children, as in adults, nothing characteristic is met with. Congestion is often present; occasionally the bones of the skull are thickened or diseased. Induration or softening of the brain is sometimes found in long-standing cases.

The prognosis will be founded, not on the violence of the fits, nor even on their quick return during a certain limited time, but on their recurrence when dentition and other palpable sources of irritation have ceased, and when the child is otherwise in good health.

Treatment during the fit consists in loosening everything round the neck, freely admitting fresh air, raising the head, and putting a piece of cork between the teeth; cold affusion to the head may sometimes be useful. In the intervals the diet must be plain and nourishing; meat must be rather sparingly given. Exercise in the fresh air should be freely encouraged, and we may administer, and hope something from, the salts of silver and iron. Belladonna is highly spoken of by Dr. West and by Brown-Séquard; Trousseau recommends that it should be persevered with for two, three, or more years, if good results. The bromides of ammonium and potassium in large doses, three, five, and ten grains, thrice daily, are of undoubted value in many cases. Children often improve remarkably under their use. Steel, quinine, and cod-liver oil are useful, in strengthening the constitution.

17. Paralysis

In childhood is not the same formidable affection as when it attacks the adult, for while its effects may indeed be permanent,—*quoad* the limb or part affected,—yet it rarely threatens life, and more frequently under judicious treatment the child is completely restored. Paralysis is occasionally congenital, however, in which case it is of course not susceptible of relief. The causes of paralysis in the child are far more often comparatively slight, *e.g.*, dentition, constipation, cold, slight pyrexial attacks, &c.; it is commonest for one leg to be affected, sometimes both are, and sometimes one arm and a leg; one arm is seldom affected alone. Sometimes the mischief is confined to a single muscle, *e.g.*, the sterno-mastoid. Occasionally attacks of paralysis occur which are dependent on very serious organic lesions, *e.g.*, chronic hydrocephalus, brain abscess, tubercular meningitis, and various tumors of the brain and spinal marrow; one remarkable feature in such attacks is the suddenness of the invasion. Dr. West points out that cases occur in which the paralysis is not complete, and in which hyperæsthesia exists for a time, to be followed by more complete loss of motor power afterwards. In such cases the child stands on the healthy limb, turns the foot of the affected side inwards when walking, and stands with the toes of that foot on the dorsum of the foot of the sound side. Hence arises a question of diagnosis from hip-joint disease, which may be solved by noting the hyperæsthesia of the paralyzed limb varies greatly at different times, while the pain in the knee is absent, and no acute pain is caused by striking the head of the femur against the acetabulum by a blow on the heel. Moreover, the temperature over the joint supposed to be affected will aid in the diagnosis, such temperature being increased on inflammation (Hilton).

The milder forms of paralysis may be diagnosed from the more severe forms by attention to the history of the case, the presence or absence of nervous tremor, twitchings or contraction of the fingers and toes. If convulsions have preceded the attack, simple paralysis is usually preceded by one fit, while brain mischief causes a succession, during which the limb to be palsied suffers peculiar movements, or may be the only part in which movement occurs. The duration of simple paralysis in the child varies greatly, some cases recover in a few weeks, while others last for years. When the duration is long the paralyzed limb wastes and declines in temperature, and its growth is also retarded.

The temperature may fall from two to ten degrees, the muscles atrophy, and fatty degeneration also occurs, and this wasting and palsy of the muscles is accompanied with relaxation of the ligaments, which causes many of the deformities of children, for instance the varieties of club-foot, &c. Palsied limbs may begin to atrophy in four or five weeks, more often months elapse first.

It is remarkable that the deltoid and *tibialis anticus* muscles waste more rapidly than any others. It is also remarkable that when atrophied muscles have entirely lost their power of contracting to the most powerful induced electrical currents, they will still react vigorously to a direct (galvanic) current of low tension and slowly interrupted (Netten Radcliffe). So long as either current produces reaction cure is eventually certain, and it is singular that as the muscles regain tone the power of the direct current lowers, and that of the induced current rises; when fatty degeneration has commenced cure is well-nigh hopeless. Duchenne recommends a small "Emporte pièce" to examine morsels of muscles under the microscope, and the current should be direct or induced according to the condition discovered, which may be further assisted by remembering the period of the case. Instruments of Stöhrer of Dresden appropriate to this disease are kept by Pratt, of Oxford Street.

So-called atrophic infantile paralysis occurs chiefly between the ages of six months to two years, and dentition is a prominent factor in its production. It is usually primary and occurs suddenly in the midst of good health; occasionally, however, there are some slight prodromata as feverishness and pains in the back and loins.

Progressive myosclerotic paralysis or paralysis with apparent hypertrophy of the muscles is very rare. It has always occurred in boys and oftenest in infancy, about twelve to fourteen months. The muscles of the legs are those usually first affected. There is a steadily increasing inability to walk, frequent tumbles occur. The feet are generally kept widely apart, and "lateral balancings" are, according to Duchenne, symptomatic. Another symptom generally noticed before the swelling of the muscle, is a difficulty in bringing the heel to the ground. Talipes equinus or equino-varus is the result, and there is a peculiar curvature of the spine, in which the shoulders are carried backwards, extending beyond the perpendicular of the sacrum. This in turn produces a deep curvature in the lumbar-sacral region, which Duchenne calls "saddle back" ("en sellure"). The affected muscles next swell and enlarge, especially the calves and buttocks, and the skin over them may be red, the temperature is, however, lowered, and the power of voluntary contraction is impaired and generally also the electro-muscular contractability. The sphincters and involuntary muscles are not affected. The appetite and digestion remain good, and the mind clear even to the end. After a long time varying from five to fifteen years the child is kept to bed, the paralysis increases; the upper limbs become involved; and there is no longer increase in size, but rapid diminution and atrophy of all the muscles affected. Death occurs often by some mild intercurrent affection. The pathological change of greatest consequence in this singular affection is great hypertrophy of the areolar tissue with abundant formation of wavy fibrous tissue and fat. No organic lesions in brain or cord have been discovered, and all treatment appears equally nugatory, certainly after the early stage. Electricity whether as the continuous, or induced current, and tonics are the means recommended.

Treatment.—The general treatment of paralysis will depend in a great degree on the cause of the attack; in fact the first efforts, if the child be seen early, should always be directed against the supposed cause, the relief of which often materially relieves the paralysis which was secondary to it. Thus dentition, worms, constipation, &c., must be inquired for and attended to. Purgatives will almost always be needed, as constipation commonly exists in all classes of cases; some tonic must also be given, steel answers very well in many cases. Dr. West gives the eighth part of a grain of spirituous extract of nux vomica to a child four years old, three times a day, increasing the dose till it reaches the third of a grain. A combination of quinine and steel is often useful. For a child two years old five drops of the fluid extract of ergot may be tried two or three times a day. It is very important that the limbs affected should be exercised every day by a baby-jumper, go-cart, or other such means; friction ought also to be employed. I have found the bay-salt water-bath followed by friction with a rough towel very valuable, but the remedy which has in my hands always succeeded best is electricity, the continuous current being passed at first two or three times a week, and gradually more frequently along the course of the nerves. It should be done for ten minutes or a quarter of an hour at a time, and the current may be strengthened gradually as improvement takes place. As a certain point or stage in improvement is reached, benefit will be obtained more rapidly by changing to the induced or Faradic current.

Facial paralysis, or that of the portio dura, also occurs during dentition; it generally proves but temporary, and it involves no danger.

Facial hemiplegia sometimes occurs in infants soon after birth. It is possibly caused by some injury to the seventh pair by the forceps, or pressure of the head in delivery. It usually rapidly diminishes and requires no treatment.

18. OTORRHŒA.

A discharge from the ears is exceedingly common in children; it proceeds from subacute catarrhal inflammation either of the external meatus, or of the membrana tympani, or of the mucous membrane lining the tympanum. The last condition is often complicated with rupture of the membrana tympani, through which the matter has made a way of egress, and, indeed, this accident is in the direction of safety, inasmuch as if the matter can find no outward way of escape it will burrow inwardly and lead to disease of the brain or its membranes. Abscess in the cerebrum is caused occasionally by disease originating in the mastoid cells, when the matter has been unable from some cause to find outlet. Earache, discharge from the ears, and deafness, are symptoms always requiring attention, lest the brain should become affected; sudden cessation of the discharge is often the immediate prodroma of such an occurence. Should such occur warm fomentations and perhaps a leech or two will aid in reducing the inflammatory action. When the disease is strumous and chronic the treatment has already been given under the head of scrofulosis. Acute symptoms, such as convulsions, vomiting, delirium, and strabismus, indicate only too surely that actual cerebral mischief has taken place; the case is then very serious, but something may be hoped from counter-irritation behind the mastoid process. A few leeches over the side of the head if there be much pain and tenderness to the touch, warm applications, especially steam, poppy fomentations, light diet, and a darkened and silent room, for great excitability and intolerance of light, are not uncommon features in such cases. The bowels will of course require attention, and sedatives are indicated if the pain and restlessness are very great.

19. CEPHALHÆMATOMA

is a sanguineous tumor developed under the scalp, under the pericranium, —or within the skull; of these the subcranial is the rarest and most intractable.

These tumors may vary in size from a walnut to an apple. They are often the result of pressure from protracted labor. The swelling is circumscribed, soft, and fluctuating, and generally disappears by absorption in a few days. It can, therefore, rarely be necessary to lay it open unless inflammation and suppuration occur, when it should be treated on general principles.

There can be no harm in the use of a lotion of hydrochlorate of ammonia as recommended by some writers; others prefer spirit lotion.

Setons have been successfully used in a few cases.

20. DISEASES OF THE SPINAL CORD.

Irritation and congestion are not altogether uncommon. Some years ago I saw a child whose gait had become tottering and unsteady, and who walked with shuffling, dragging movement of the legs, who showed, moreover, great listlessness and unwillingness to walk, but who was fat, and ate and drank heartily. This child was addicted to the practice of masturbation to a frightful extent. Blistering the penis broke the habit considerably, and by unremitting vigilance the child was at length cured of it, and when I again saw him, two years subsequently, he was quite well and able to walk and run properly.

Inflammation of the cord is rare in this country; it has, however, prevailed epidemically, especially in Ireland and France. Boys of about twelve years of age were commonly attacked by it. The invasion of the disease was sudden, with abdominal pain, purging, and collapse, followed in a few hours by burning heat, stiffness of the muscles, retraction of the head, convulsions, coma, and death. The spinal arachnoid was the part chiefly affected, but the brain membranes were also involved. There was generally considerable effusion of lymph between the membranes of the cord, the cord substance itself being healthy or a little congested.

The ordinary sporadic cases of inflammation of the membranes of the cord are rare, and a very brief description will suffice. The bowels are constipated; often obstinately so. There is slight headache and vomiting; often shooting pains about the limbs and back. Then all degrees of loss of motor power may be present from a mere dragging or weakness of the legs up to complete paralysis; stiffness and spasm are regarded as especially indicating spinal meningitis, and paralysis as pointing to inflammation of the cord substance.

Usually tenderness is felt on making pressure at some portion of the spinal cord. There is often freedom from delirium and intolerance of light; but the vomiting and pains in the back persist; sometimes severe convulsions come on, the head retracted, and a condition of opisthotonos produced. The vomiting is then more urgent, the pulse becomes very rapid, and convulsions soon end the sufferings.

Chronic inflammation of the substance of the cord is generally due to caries of some of the vertebræ, or to mechanical injury from blows, strains, &c., without absolute disease in the bones. Strumous children are the usual subjects of it. In either case paralysis is the result, but in caries this is due not only to the inflammatory action, but also to the spinal distortion and the pressure on the cord from displacement.

Treatment in affections of the spinal cord involves rest; the lateral or even prone decubitus being chosen according to circumstances. The spinal ice bag in acute affections, with small and repeated doses of aconite, or aconite and gelsemin, is preferable to all lowering measures whatsoever. In chronic cases, attention must be paid to the bladder and bowels, the little patient being kept strictly clean and dry. Nutritious but stimulating diet and general hygienic rules must be enjoined. Of medicines, the tincture of gelsemin is really valuable in spinal irritation. I recommend its use in small doses in neuralgia, cramps, spasm, nervous irritability, and wakefulness, frequently with excellent results. It is a remedy that will repay further study in its action in affections of the nervous system. Conium is another medicine used in chronic spinal affections; there is some doubt of its efficacy. Iodide of potassium and bichloride of mercury are to be used if the case be syphilitic, which, I am persuaded, they frequently are; and bromide of potassium may also be usefully employed.

I have quite recently (August, 1876) seen a desperate case of syphilitic spinal and cerebral disease recover. The patient was a young doctor. The vomiting, deafness from otorrhœa, violent convulsions, obstinate costiveness, and other features of this case, rendered it so formidable that several medical men, who saw it in consultation with me, agreed with myself that it was all but hopeless. Nevertheless, ice, iodide of potassium in very large doses, and, occasionally, bromide of potassium, enemas, and nutritious diet, pulled him through, and he reports himself to me in August as "at last after four months' illness able to write, improving rapidly though still weak in the lower extremities; deafness worse; right side of face rather numb, especially when the temperature is low; constipation still troublesome." This gentleman then took iodide and bromide of potassium with a small quantity of perchloride of mercury, and tincture of nux vomica. He also used Faradization over spine and abdomen. By December, 1876, he was attending to his practice, and better than for a considerable time previously. I refer to this case merely as showing the value of treatment in even very desperate cases, and where one might think the nervous centres irretrievably damaged.

21. Trismus.

Infantile trismus, or tetanus neonatorum, or nine-day fits, is a disease hardly, if ever, seen in this country; it was once common, but has vanished before cleanliness and improved hygienic conditions. It is common still in the West Indies, Germany, Minorca, Ireland, &c.; it may come on within twelve hours after birth, or not for some days; it is rare after the week. The child is observed not to take the breast, it cries, and the mouth is found to be somewhat fixed; the infant is seized with violent irregular convulsions, with foaming at the mouth, thumbs riveted into the palms of the hands, the jaws locked, the face and even the whole body becomes livid, and the disease may be fatal in from eight to forty hours.

Post-mortem.—There is generally found an effusion of blood, fluid or coagulated, into the cellular tissue surrounding the theca of the cord; the vessels of the spinal arachnoid are usually congested, and sometimes there is an effusion of blood or serum into its cavity. Different writers have attributed trismus to a great variety of causes. Probably Dr. Joseph Clarke is right in attributing it to impure air, to neglect in keeping the infant clean and dry, and to the abuse of spirituous liquors on the part of the mother; and these views are confirmed by the variety of places widely apart as the poles where it occurs, under every variety of climate, under every different method of dressing and treating the umbilical cord, and in fact where the only conditions in common are improper diet, bad smells, and neglectful management. For instance, in the year 1782 one-sixth of the children born in the Dublin Lying-in Hospital died within a fortnight of birth, and nineteen-twentieths of their deaths were caused by trismus. But out of 16,654 infants born during Dr. Collins's mastership, only thirty-seven died of trismus since the improved ventilation and other hygienic conditions. While prophylaxis is thus successful, treatment is unfortunately quite the reverse, every description having been tried and found useless. Leeches, blistering, and hot baths, opium, dressing the cord with turpentine, Indian hemp, enemas of tobacco, calomel, musk, and ambergris, tincture of iodine and assafœtida; all have been tried, all have their advocates, and all seem to be useless in the majority of cases. Whether or not the umbilicus is ever the cause of the disease by taking on unhealthy ulceration is also a moot point. Dr. Colles maintains this to be the essential cause of trismus. Dr. Labatt and Dr. Breen entirely contradict this view. It is probable, however, that more than one factor may give rise to the disease, some conditions being more prevalent in one class of cases, and others in others. At least only in such a way can we reconcile the opposite opinions and observations of conscientious and competent authorities.

22. Hydrorachis, or Spina Bifida,

is a congenital malformation, consisting of one or more tumors in the spinal column in its lumbar, dorsal, or cervical portions, which communicate with the spinal canal. These tumors vary considerably in size; they are sometimes semi-transparent, and the skin covering them may be natural in appearance or inflamed or bluish, almost livid. The disease is caused either by an arrest of development in some portion of the spinal canal, or by an excessive production of the cerebro-spinal fluid, which by pressure causes absorption of the bone and ultimately protrudes in the manner described. The commonest site is over the lower lumbar vertebræ. There is usually an absence of a greater or less portion of the lateral arches of the vertebræ, the spinous processes being divided or absent. It happens sometimes that the outer covering of the tumor is not skin, but dura mater, pia mater, and arachnoid only. Cases have been recorded in which the dura mater was absent, and the other membranes formed the only covering.

The fluid contained in the tumor is generally clear serum; it is sometimes, however, sanguinolent and even purulent. The quantity varies from one to seven pints. The spinal cord traversing the tumor is often unaltered; sometimes, however, it is lengthened, or flattened, or destroyed, into shreds or filaments.

Pressure on the tumor causes general uneasiness, and if persisted in convulsions and coma. The symptoms produced by this disease are want of power in the lower extremities, often the limbs are atrophied and paralyzed. Sometimes no abnormal conditions exist. These differences doubtless depend on the amount of pressure exerted on the cord. The tumor sometimes bursts before birth, without necessarily killing the child. More often, however, inflammation and ulceration take place and the tumor bursts, ending rapidly in convulsions and death. The bursting of the tumor generally takes place before the third year is attained.

Treatment is far from being satisfactory; in fact, many writers have come to the conclusion that the more the tumor is let alone the better. Various operations have, however, been proposed and practised; probably the best is puncturing with a fine trocar well at the side of the tumor to avoid the spinal cord, and afterwards maintaining equable pressure to promote absorption. Iodine injections have been also employed, and in one or two cases with success. Pressure alone has also been tried. In any case the general health should be attended to; the child must be strengthened as much as possible, and instances have occurred in which persons have grown up and attained considerable ages without any accident occurring to the tumor.

CHAPTER VII.

DISEASES OF THE AIR-PASSAGES AND THORACIC ORGANS.

1. CORYZA

is an affection singularly common and of great importance during the first two months of existence. The vulgar name for it is the snuffles. It is ushered in with slight feverishness and a snuffling sound in breathing, and sneezing. At first there is but little discharge from the nostrils, afterwards this becomes abundant, sometimes acrid and even muco-purulent. It then dries and forms a crust about the nostril which greatly impedes the child's respiration, and causes it to breathe through the mouth. If the nasal breathing is completely obstructed the child will be unable to suck, because were it to do so it must be suffocated. It will then require to be fed with the spoon. The possibility of snuffles consequent upon syphilitic disease is always to be borne in mind (see Syphilis). Occasionally a common cold is but the prelude to some severe attack of disease, *e.g.*, measles, &c. Cold in young infants is often caught by imprudent washing, undue exposure, &c. The treatment will comprise small doses of spirits of camphor, mild diaphoretics, perhaps an aperient, warmth, and the removal of the crusts. The formation of the crusts is best prevented by smearing the nostrils with a little weak bismuth ointment and using great cleanliness. Dr. Meigs recommends that the interior of the nostril be touched with some animal oil or cucumber ointment and a loose flannel cap be worn on the infant's head. In very severe cases the interior of the nostril may be touched with a solution of nitrate of silver (gr. v.— ʒj.) several times in the day. Chronic coryza in an infant is almost always syphilitic.

For ordinary cold-catching in older children I find no treatment so useful as small doses of spirits of camphor, frequently repeated, say every hour or half hour; the frontal fulness; the running at the nose, and other well-known symptoms which lead people to say "I know I have caught a cold," will disappear, and the cold be nipped in the bud, or if the cold be more advanced than this, and what is called "feverish," with hot skin, frequent sneezing, &c., small doses of tincture of aconite, say half a drop in a little water every half hour, will speedily cause diaphoresis, diminution of fever, and great general feeling of relief. The lassitude left after a cold, or a cold showing tendency to become chronic, are best met by tincture of nux vomica. For a child, one or two, or four or five drops, according to age, three times a day, in a little water. I have put these plans of treatment so frequently and so thoroughly to the test, that I have great confidence in them. The cold bath, whether sponge, shower, or plunge, is a most useful means of preventing the liability to take cold, and infants and young children should be taught to breathe through the nostrils, by gently closing their mouths during sleep, and so getting warmed air, and not cold, to their lungs. Simple as such a procedure may appear it is a very effective preventer of colds, and the advice "shut your mouth" has much force and wisdom in it.

2. DIPHTHERIA

is an acute, specific, infectious, and contagious disease, the essential characteristic of which is a spreading asthenic inflammation of the mucous membrane of the pharynx attended by exudation of lymph, and the child may die of the general or of the local disease. The primary seat of the disease may be the mucous membrane of the *tonsils*, palate, uvula, or nares. At first there appears redness, then white or gray patches are seen from layers of lymph. Sometimes two or three centres of lymph may be seen from the very first. The disease may spread to the posterior nares, anterior nares, epiglottis, larynx, trachea, bronchi, œsophagus, and stomach; so wounds and excoriations, and even the vagina and anus, may be covered with patches of false membrane. If the lymph be removed the surface beneath is seen to be red and bleeding, and in a few hours becomes covered with a fresh layer of lymph. This effused lymph may be granular and creamy, or very tough, like washed leather; it is elastic, and sometimes as much as one-eighth of an inch in thickness. The exudation matter consists of epithelial and granular cells, with granules of fat and protein. Sometimes fibres and vegetable growths are present.

Diphtheria varies much in its general characters. Sometimes the general and local symptoms are trifling, attended with little fever, little soreness of the throat, slight dysphagia, and no nervous symptoms, nor yet any albumen in the urine. Sometimes again the disease is terribly severe; the mucous membrane may be vivid red or livid, the dysphagia extreme, the fever great, but of low adynamic type, intense muscular weakness, the urine loaded with lithates, albumen, and granular and sometimes tubular casts. Again, the general symptoms may be slight, the laryngeal symptoms severe, insidious, and even fatal in a few hours. In cases of this kind there is croupy cough, altered voice, and suffocation imminent from the first. In children who have had scarlatina there is often noticed a sanious discharge from the nose, some tenderness and swelling at the angle of the jaw, and on examination diphtheritic exudation is found on the fauces. This is, however, not true diphtheria according to some writers. Sir John Rose Cormack ("Edin. Med. Journal," March, 1876) thus contrasts the pellicle of diphtheria with that of scarlatinal sore throat. That of diphtheria is a tough strong membrane difficult to tear, the scarlatinal simulation being a pultaceous easily torn stratum of detritus, or a portion of gangrenous mucous membrane. Nevertheless, Sir John goes on to remark, "there exists in some cases a great difficulty in establishing an absolute diagnosis. Diphtheria, as Graves and others have shown, and as Trousseau has admitted in his later teaching, *often follow so close in the wake of scarlatina as to seem one of its later stages.* False membranes, diphtheritic membranes form in the pharynx and larynx, as an immediate sequel of scarlatina. In such cases diphtheria has closely followed, or been engrafted in the original attack of scarlatina."

The italics are mine, and I confess one is driven to feel with the late, justly eminent Dr. Tracy of Melbourne, "I think the distinction between a bad scarlatina sore throat with exudations, and diphtheritic sore throat, is very fine indeed." Death may occur from apnœa or asthenia, or by pyæmia. The mind is usually clear; vomiting and delirium if long continued are very fatal symptoms. So also is great scantiness or actual suppression of urine. Albuminuria is always a grave symptom, hæmorrhage also, and the blood is generally fluid, so much so that death has been reported as following leech bites.

In 1872 a report was submitted to the Legislative Assembly in Victoria, embodying the experience derived from an immensity of evidence (as well as the evidence itself) taken on the subject. From this report I have ventured to compile the following points, which were considered established:

1. The period of incubation is three or four days.
2. Diphtheria may occur in any locality, and under every condition.
3. It attacks indiscriminately rich and poor, the well-fed, and poorly nourished. Except that a porous soil, with an understratum of clay, appeared rather to favor its development, geological strata appear to have no relation to its prevalence.
4. While bad hygienic conditions of necessity promote its virulence by lowering vitality, &c., good hygienic conditions afford no exemption from its attack.
5. Temperature has little effect, the moist cold weather preceding the winter was the time of its greatest prevalence.
6. Even diathesis seems to make little difference, though some of the medical witnesses examined considered struma as a predisposing cause.
7. Children from three to eight years are the most frequent subjects. Children at the breast rarely suffer.
8. The duration in favorable cases is from ten to fifteen days. Termination in death or in recovery generally becomes apparent in seven days.
9. Recurrence in the same individual is not the rule, and recurrent cases are less severe than the original attack.
10. As a sufficiently singular fact the public institutions (with one exception) were remarkably free, and the *Chinese* race seemed to possess absolute immunity.

Sequelæ.—Great depression of the pulse; nervous debility; paralysis, especially about the soft palate and the pharynx, but occurring also elsewhere, and affecting sometimes the limbs. Impairment of vision,—this is rarely of long duration.

These paralytic symptoms are not apparently predisposed to by the gravity of the preceding attack, but, like the dropsy following scarlet fever, may come on after cases so slight as scarcely to attract attention. Nor is the presence, absence, or quantity of albumen in the urine a better guide. The paralytic symptoms may come on shortly, or several weeks after, the cessation of the primary disease.

The paralysis may be motor or sensory or both, it is so far as the extremities are concerned never strictly unilateral, it is progressive and gradual, and the mind is seldom affected. The mean duration of the paralytic symptom is about a month, but the time may vary from a fortnight to several months. Of seventy-seven cases collected by Dr. Russell Reynolds only nine were fatal.

Guersent has pointed out that diphtheria is especially liable to complication with a low insidious form of pneumonia, a pneumonia of crasis of the blood presumably, a pneumonia of stagnation, which requires the free use of stimulants and carbonate of ammonia to meet it. Huskiness with short cough and expectoration of mucus streaked with blood and frequent respiration are signs indicative of this condition. The stethoscope will complete the diagnosis.

There is a family likeness between diphtheria, croup, scarlatina and idiopathic erysipelas. Between croup and diphtheria the following distinctions are recognizable, the character of the inflammation is in croup sthenic, in diphtheria asthenic; in croup the site of the inflammation is at first confined to the larynx extending thence to the trachea and bronchi, whereas in diphtheria it is the tonsils, pharynx, soft palate, extending thence to the nares and air passages; besides which in diphtheria there is fetid breath, a sanious discharge from the nostrils, and the neighboring lymphatic glands are swollen, which symptoms are absent in croup, and there is besides marked difference in the sequelæ of the two diseases. Croup is, moreover, as a rule, sporadic, diphtheria as a rule epidemic (see Croup).

From idiopathic erysipelas of the throat (a rare disease) the diagnosis is not easy, but in erysipelas there will be more puffiness and swelling of the throat and surrounding parts, which are besides often œdematous; the exudation is less membraniform, and the tongue dark brown, or black, dry, and deeply fissured.

The following are the chief diagnostic points between diphtheria and scarlatina:

1. An attack of either disease confers no immunity from an attack of the other.

2. The peculiar rash of scarlatina is absent in diphtheria, though there may be in the latter an erythematous blush, evanescent and occurring in patches, but this is by no means a constant symptom.

3. Albuminuria occurs in the course of diphtheria, but towards the close of scarlatina.

4. The strawberry tongue of scarlatina and the subsequent desquamation are also points wanting in diphtheria.

5. The sequelæ are totally different in the two diseases.

Treatment.—Remembering that the contagious elements unquestionably exist in the breath, in the exudations on the throat, and elsewhere, and also probably in the excretions, disinfectant means must be put in force. The room must be well ventilated, fresh air is an absolute necessity to oxidize the blood, and to prevent the accumulation of the poison. Dr. Reynolds, of Mansfield, relates two desperate cases in which, as a *last resource*, he had the children carried about in the open air, night and day, with their mouths open, and as much as possible facing the wind. Both patients *recovered*. I recommend that the room be atomized with sulphurous acid vapor, and that sulphurous acid be dashed upon sheets hanging across the door, so as to disinfect the air passing into the house. I would recommend all persons passing from the sick to the healthy, to fumigate themselves and their clothes with sulphur vapor. All hangings, carpets, &c., are to be removed from the sick room. All excretions are to be passed into vessels containing carbolic acid or other disinfectant. The nasal discharge, especially infectious (also a serious symptom), must be cleared away by antiseptic injections: a proper continuous stream is essential to do this properly—a glass syringe is mere playing at it. Condy's fluid, or a lotion containing hypochlorite of soda, may be used for the nasal douche. If we cannot rely upon the attendants being unremitting with the use of the sulphurous acid apparatus, it is better to have sulphur burned every four hours on the live coals or a hot shovel; it should be burned for half an hour at a time. In the second edition of this book I stated that carbolic acid and glycerine was an application of which I had then no experience, but which *would probably be valuable*. It is hard to over-estimate the value. Carbolic acid, just made fluid with a little water —but glycerine is even better, and applied topically forms perhaps the most useful local application we possess; whether the action be caustic, or disinfectant, or a little of both, matters not, so that the result be good, as it unquestionably is, on the testimony of many medical men in Victoria, New South Wales, and elsewhere, where the experience of diphtheria is great. I regard as next most useful strong tincture of iodine topically applied. To the value of this there is also abundant testimony. Various writers recommend a solution of nitrate of silver. Sir W. Jenner says, one thorough application, of the strength of one scruple to a drachm of water (Ðj. to ʒj.), should be painted over with a brush. Then Tinct. Ferri Perchloridi, and equal parts of Liquor Ferri Perchlor. and glycerine, have been used, also hydrochloric and nitric acids, diluted with glycerine and water. A solution of chlorinated soda (Liq. Sodæ Chlorinat. ʒss. ad Aquæ ʒvj.) may be syringed into the throat instead of a gargle, as, of course, young children cannot gargle. The membranous exudation is on no account to be torn off.

Regarding constitutional treatment, the most important indication is to support the strength in every way, the disease being of perhaps more lowering character than any other with which we are acquainted; and the other indication is to control, and if possible prevent, the formation and spread of the false membrane. Three remedies have been especially commended in the general treatment of this disease—chlorate of potash, hydrochloric acid, and tincture of steel. Probably a combination such as

℞ Potassæ Chloratis.................................... gr. x.
   Tinct. Ferri Perchlor................................. ℔v—x.
   Glycerini... ʒss.
   Aquæ... ʒij.
Ft. Mist.

is the best. At the same time, beef tea, soups, wine, and every form of light nutriment may be given with advantage. Rennet whey and buttermilk are recommended as a drink by Dr. Condie. The quantity of wine and brandy absolutely needed is sometimes very large, and, indeed, can hardly be overdone. When dysphagia renders swallowing impossible, enemata of port wine and beef tea must be resorted to. In obstinate vomiting, which is so dangerous a symptom, ice may be sucked, while the strength is supported by the rectum, and if the sickness does not pass away, the hypodermic injection of a little morphia over the stomach may be tried. When albuminuria is a primary symptom the tincture of steel may be omitted, and iodide of potassium substituted; and the child should, when practicable, drink freely of barley-water and other diluents.

Bichromate of potash is highly spoken of by Dr. Orton of America, $\frac{1}{16}$ to $\frac{1}{8}$ grain to be given in water every half hour till vomiting is caused, when it is to be continued every two hours. Dr. Orton states that this treatment was successful with a single exception in 142 cases. Dr. Maynard recommends hyposulphite of soda as a local application (ʒij. with ʒij. of glycerine and ʒvj. of water), to be applied two or three times a day to the throat. One to three grains may also be given internally to young children every four hours. Lemon juice is recommended as a gargle by Dr. Revillent, and it has this advantage that it is of benefit rather than otherwise if some be swallowed.

Ice is always valuable and may be freely sucked. The inhalation of iodine is also advocated. All attempts at bleeding, blistering, or indeed any form of counter-irritation, are worse than useless. When the larynx is involved and dyspnœa very urgent, tracheotomy (in children rather laryngotomy) has been successful. It is, of course, a forlorn hope, but should be undertaken when the lungs are in favorable condition. Care must be taken not to push the canula between the false membrane and trachea. During convalescence, rest, quinine and iron, and lastly, strychnine is the tonic from the use of which much may be hoped for in the prevention and cure of any remaining paralysis, and galvanism is the most useful local agency in these cases. Change of air is also desirable in restoring the enfeebled heart and nerves.

3. Croup—Cynanche Trachealis—Acute Membranous Laryngitis.

This disease is very alarming in its symptoms, slightly if at all contagious, but very fatal; it therefore deserves a careful consideration. It is most common during the second year of life, rare after the fifth. It is more common in boys than in girls. It is rare for genuine croup to recur in the same individual; the attacks of so-called croup from which some children are stated to suffer year after year are not real croup at all. When it does so recur the second attack is usually milder than the first. It has, indeed, become a serious and hotly discussed question if the disease croup is to be retained at all as having any separate existence from diphtheria. For more than six months in 1875, was battle waged in the columns of the "Lancet" upon this matter, and despite the high authority of those who urge their identity, I confess to remaining unconvinced. Sir W. Jenner speaks of the identity of the exudation, that there are no anatomical characters by which it can be distinguished in the two diseases. But what anatomical character is there to distinguish gonorrhœa from ordinary vaginitis, and yet the one arises from a definite and specific cause, and the other arises idiopathically beyond doubt. And we have even greater distinctions in character, course, and sequelæ, between croup and diphtheria than could be asserted to exist between gonorrhœa and ordinary acute vaginitis in a chaste woman. But what good purpose is served by asserting the identity of these? I think the admirable exposition of Dr. Thomas, of New York, upon this point worth quoting here. He says, "as there is but one cause for scarlet fever, for measles, and for variola, namely, absorption of a specific poison or contagious material, so is there, it appears to, but one cause for gonorrhœa. It is true that simple acute vaginitis may simulate gonorrhœa so closely that the experienced observer will be foiled in diagnosis, but this does not prove the diseases to be identical. The poison of gonorrhœa produces inflammatory results as a certain consequence of contact; the causes of acute vaginitis produces them as an accident, which probably in a different state of the patient's system would not have occurred." The poison of diphtheria, and the idiopathic origin of croup, seem to me a similar example.

But the distinctions between croup and diphtheria are, I think, sufficiently definite beyond this. Similarity of cases here and there does not remove the great broad line which, looking to the totality of the symptoms, appears to me to run plainly enough between the two. In diphtheria I recognize a disease epidemic and eminently contagious, of an intensely asthenic type from its very commencement, in which exudation is formed upon the tonsils and pharynx, and spreads thence, upwards and downwards, with a disposition to form upon distant surfaces (*e.g.*, a blistered surface), which occurs at all ages, and in which paralysis is a common sequela; whereas in croup, I recognize a disease sporadic, and doubtfully if at all contagious (Sir Thomas Watson, in 1857, teaches distinctly that it is not contagious, and speaks of diphtheria " as this very formidable complaint of which I have *not seen more than two or three examples* "), of a rather sthenic character at first, in which *not* the tonsils and pharynx, but the larynx and trachea, are the parts first attacked, in which the tendency to spreading is far less marked, which is exclusively a disease of childhood, and of which paralysis is not a sequela. Any one keeping the two types in mind, however near special cases may approach the border, will scarcely fail to recognize two diseases as distinct as the ordinary types of measles and scarlatina. In the words of Dr. Wilks, " our knowledge at the present day seems to show that a membranous inflammation of the air passages may be caused by injury, by cold as an idiopathic affection, and by the diphtheritic poison." The invasion of the disease is sometimes sudden, more often it is preceded by a cold—some feverishness, thirst, hoarseness, and running at the nose; slight sore throat is also complained of, or if the child be too young to speak, it puts its hand roughly to its throat or rubs it. This, the so-called first stage, may last from twenty-four to thirty-six hours, when it passes into the second stage, which is especially marked by the cough becoming " brassy " or "clangey." The act of inspiration being prolonged, the breathing is hard and attended with a peculiar sound, which, like the cough, is characteristic of the disease. This sound is crowing or barking in character, and once heard will be remembered. Dr. Meigs compares the cough of the croup when it has covered the interior of the larynx to the sneezing of a young kitten. The paroxysms of cough and dyspnœa get always worse (as in the case with most of the severe symptoms in all diseases of children) towards night; in the morning there is remission, which is natural, and which must not deceive the physician as to the real state of the child. By this time the fever is increased, the thirst is greater, the tongue is furred; the child is restless, and, in fact, as the disease proceeds, is constantly fighting for breath. At first, between the paroxysms, he would drop into an unquiet sleep from sheer exhaustion; but later on there is no time for this. The fits are almost incessant, the face is most anxious, the eyes glassy, the lips livid, and clammy perspirations begin to break out. The child throws back its head to increase the size and capacity of the trachea to receive air. The child, if able to speak, says little from the pain of speaking, but swallowing is usually well performed. The case may now go on from bad to worse, suffocation becoming more and more imminent, until the patient sinks in coma, or convulsions come on and end the struggle.

More often, however, the end is not just yet; the child seems better, perhaps it has been relieved by treatment. Many of the worse symptoms may seem less urgent; some expectoration may take place and the child appear altogether better. But in a few hours the former troubles may return, like armed men who have been but resting themselves for a final effort, which then ends rapidly in death; or the child may pass by slower and more imperceptible stages into a comatose state from which his nurse and attendants find suddenly that it is impossible to rouse him.

The auscultatory signs in the second and third stages are chiefly a weak respiratory murmur in the chest, which nevertheless sounds well on percussion. Moreover, there is a concave state of the intercostal spaces at each inspiration. Sometimes from the first there is a diffused sonorous *râle* indicative of bronchitis. Should pneumonia set in there will be the small crepitation belonging to it, with distinct dulness on percussion over the inflamed portion of lung. This pneumonia is usually double. These sounds will be, of course, masked to some degree by the croupy noise in the trachea.

The duration of an attack of croup varies with the strength of the patient, and the intensity of the inflammation; the average is from two to five days. Sometimes the disease terminates in fourteen hours from the commencement. Dr. Cheyne says that it usually proves fatal on the third, fourth, or fifth day.

Post-mortem.—The mucous membrane of the larynx and trachea is usually inflamed, red, vascular, thickened, and peels off easily; the characteristic materies morbi is the membrane which lines the air passages. Microscopically this consists of cells in abundance, lying in finely fibrillated substances. It is a layer of lymph of variable thickness, white, yellow, or ash-colored, lining the larynx and trachea, and extending down even into the bronchi. This false membrane may be in patches or in cylindrical pieces, or present perfect moulds of the tubes. It is thinner in the larynx than in the trachea, and thinnest in the bronchial tubes. Its free surface is smooth and often glazed with muco-puriform matter; the other surface is more or less adherent to the mucous membrane. The bronchi show traces of inflammation throughout their extent. Lobar and lobular pneumonia not unfrequently exist, and vesicular emphysema is generally present at some portion of the lung. The right side of the heart is generally gorged with dark blood. Congestion more or less may exist in the brain, liver, spleen, and kidneys. Enlarged lymphatic glands are generally found beneath the thyroid, on either side of the trachea. Occasionally the right auricle is filled with a fibrinous concretion formed during life, and marked with the currents of blood passing over it; in such cases death has taken place from the heart.

Prognosis.—Generally unfavorable, though even severe cases may and do recover; it is probable that half the children attacked die. Favorable symptoms are early and free expectoration, the breathing remaining free, the voice being little changed or recovering its natural tones, the pyrexia moderate. But if, on the other hand, there is great difficulty in breathing, high fever, much clanging noise, and no expectoration, the case will probably die.

Diagnosis from laryngismus stridulus, by the absence of fever and the freedom of breathing in the interval of the attacks. Children are apt to suffer from a form of catarrh, with so-called croupy cough, and these attacks are frequently spoken of as croup; but they may be readily distinguished by the absence of fever, the tranquil respiration, and the ready subsidence of the symptoms before simple treatment. (See Catarrhal Spasmodic Laryngitis.)

From laryngitis by the fact that that disease is far more common in adults, and causes a fixed pain in the larynx, and there is no exudation of false membrane.

Treatment.—The child must be placed in bed in a room well warmed; a warm bath will be useful at the outset, after which an emetic may be given; a teaspoonful of Vinum Ipecacuanhæ, repeated if necessary, answers well.

Dr. Meigs recommends alum in doses of ℨj. in honey or syrup as a sure, safe, and non-depressing emetic. The hypodermic injection of apomorphia (gr. 1/10 is another method which is very certain and non-depressing.

The air of the room is to be kept moist as well as warm; in a severe case a blanket tent round the bed will be needed, and the spout of a vessel in which water is boiling should send its steam into the enclosure. In milder cases it will suffice to have a kettle boiling in the room, with some contrivance to prevent its steam going up the chimney. It is a good plan to ring out sponges in hot water and to apply them near the child's throat. In severe idiopathic croup general bleeding is usually recommended. I confess to having the greatest aversion to the practice. Should the pulse run high, and the case be of full sthenic type, a dose of calomel may be required, and then I recommend one drop of tincture of aconite in a little water, every half hour for say four hours, to be continued after that or not, as the case may require. Besides this the atomizing of iodine, sulphurous acid, or lime water, into the atmosphere round the child, should be practised. A warm bath, rather cooler than the actual temperature of the child, often acts beneficially. A blanket should cover the bath, and the child be lowered into it on the blanket. Should the disease appear in the pharynx, local applications similar to those recommended in diphtheria will be advisable. Blistering is never called for, even if lung complications arise.

Dr. Meigs speaks highly of alkaline salts, especially the chlorate of potash, which he combines with the tincture of the perchloride of iron.

R Potassæ Chloratis.................................... gr. iij.—x.
Tinct. Ferri Perchlor................................. ♏ iij.—v.
Syrupi... ℨ ss.
Aquæ.. ℨ ij.
Ft. mist., tert. quaq. hor. sumend.

The tincture of iodine is a useful pigment over the throat during convalescence. Throughout the disease it will be necessary carefully to support the strength with judicious diet; if there be much difficulty in getting down the food, or if spasmodic fits be excited the food must be given per rectum.

Other remedies highly spoken of are sulphate of copper, not only as an emetic at the outset, but throughout the disease in doses of one to four grains. Sir D. Gibb praises Sanguinaria Canadensis and Dr. Condie and other American physicians recommend hydrochlorate of ammonia in three to six grain doses. And it may also be topically applied. Copaiva, thirty to sixty drops, is extolled by Dr. Lincoln.

It will be often advisable to administer a stimulant expectorant, especially when bronchitis supervenes late in the course of the disease. Such would be a combination of Ammon. Sesquicarb. and Senega, or the ammoniated tincture of valerian and ipecacuanha. The occurrence of periods of improvement must not lead to too early a withdrawal of the remedies. Such improvements are often deceptive, and require the physician to be doubly on the alert. So during convalescence care will be needed and exposure to cold should be rigorously avoided.

Lastly, the question of tracheotomy arises, and may be reduced to a few simple propositions.

1. It is not a curative measure, but a means adopted to prevent imminent suffocation.

2. It removes the mechanical obstacle to the entrance of air into the

lungs, and diminishes the spasm of the glottis, which also interferes with respiration.

3. The operation itself is not a serious one, and may be performed with the loss of about two drachms of blood.

4. The operation possibly accelerates or causes intercurrent bronchitis; this, therefore, should be carefully guarded against.

5. The operation should be undertaken early in the course of the disease, as soon as ordinary treatment has failed and suffocation is imminent; but ordinary treatment is to be persevered with after tracheotomy has been performed.

6. Cases in which the operation is performed should be well selected; chest complications in particular should be ascertained, and if of serious nature the operation should not be performed.

4. Tracheotomy.

The following are a few important points about this operation:—Chloroform may be used with safety when required, which is chiefly when struggling is going on, and when the dyspnœa is recent and spasmodic, and the child vigorous. Hæmorrhage is to be avoided by operating with deliberation; feeling for and avoiding large vessels. The operation through the crico-thyroid membrane (laryngotomy) is easier, and in some cases preferable, otherwise it is unsatisfactory, because—

1. The opening is too small to admit of breathing being well performed.

2. The integrity of the larynx and vocal apparatus is destroyed.

3. The tubes rest against the mucous membrane, which is highly sensitive, and gives rise to irritation.

4. This irritation may lead to inflammation and ulceration within the larynx, and even cause necrosis of the cartilages.

It should be remembered in the tracheal operation that the third and fourth rings of the trachea are covered by the thyroid isthmus, which may, however, be cut with impunity, but is better avoided.

Instruments required are a canula,* a small sharp scalpel, a blunt scalpel, two pairs of dissecting forceps, a pair of artery forceps, a common tenaculum, two blunt hooks, and a double blunt hook. Mr. Durham has recently recommended a form of canula of which the curved portion is jointed like the tail of a lobster; and he lays particular stress on the value of a blunt exploring trocar on which to pass the canula.

In the operation itself the operator stands on the right of the pillow on which the child is. After chloroform has been given, pillows should be placed under the shoulders and lower part of the neck so as to throw the trachea well forward. A long outer incision is then to be made (the forefinger and thumb of the left hand steadying the trachea, and restraining the violent up and down movement which often seriously embarrasses the operator); this incision should be from immediately above the cricoid cartilage down towards the sternum, for one inch or one inch and a half. The sterno-hyoid muscles here lie side by side, and should be separated

---

* These are of all sizes:

No. 1.—$\frac{3}{16}$th of an inch for a child 1 to 4 years.
    2.—$\frac{4}{16}$th   "   "   5 to 8 "
    3.—$\frac{5}{16}$th   "   "   9 to 12 "
    4.—$\frac{6}{16}$th   "   "   14 to 16 "

by assistants. The cellular tissue should be broken down with the blunt scalpel. The trachea is next to be searched for with the finger, and four rings of it exposed. The sharp hook holds the ring next above the one to be cut; this cut should be exactly in the middle line. When the trachea is opened there is a boisterous outrush of air and mucus. The tube should then be immediately inserted, a process which may be facilitated, says Mr. Heath, "by placing the handle of the scalpel at the upper end of the opening, and turning it at right angles with the trachea." To dress the wound no sutures of any kind are required; a piece of lint soaked in oil and covered with oiled silk should be placed beneath the collar of the canula, and a piece of muslin should be fixed to act as a respirator. A blanket tent must be made over the bed—the air kept at 65° to 70° Fahr. day and night, and moistened. The tube must be removed and cleansed three times a day at least, sometimes every three or four hours. The earlier the tube can be taken out the better; a cork should be put in to try the respiration, and withdrawn if difficulty of breathing is caused. This should be done daily, and left in longer and longer until the tube can evidently be safely removed.

Trousseau recommends that the edge of the wound be cauterized the first three or four days with nitrate of silver to prevent the formation of false membrane. Noisy breathing is the indication by which we know the tube needs cleansing. The inner surface of the tube may be smeared with pure glycerine. Dr. Meigs recommends the atomizing of lime-water through the tube every few hours, when the breathing seems noisy. The atomization excites cough and softens the viscid mucus which the child then rejects through the tube. He considers that the solvent action of lime upon the exudation assists in this result. Sometimes a piece of false membrane too large to escape through the canula requires to be got rid of, the tube should then be taken out and if the false membrane be then ejected the tube may be returned but not otherwise. The wound heals after the removal of the tube in about a month. It may be, however, considerably less or more. When there is dysphagia after the operation it may be sometimes overcome by the child placing its finger on the tube at the moment of deglutition. If this should fail the question of feeding becomes difficult and requires the greatest attention. The rectum must be mainly depended upon; to prevent its becoming irritable, a few drops of laudanum should be added to each enema. Sometimes the difficulty in deglutition is confined to liquids. Dysphagia is rarely noticed before the third day and seldom beyond the twelfth day after the operation. Sucking ice is a valuable means of quenching thirst in such cases, and bathing the surface with tepid water is also refreshing. Cream, vermicelli boiled in broth, eggs much cooked in milk, turtle soup, and Liebig's extract of meat, are the most concentrated nutriments, and brandy the best stimulant. Dr. B. W. Richardson has recorded a case where life was saved, when the child was apparently dead, by the use of the double-action bellows in maintaining artificial respiration, the tube having become choked.

5. Spasmodic Croup—Spurious Croup—Catarrhal Spasmodic Laryngitis.

This is an affection of by no means rare occurrence amongst children. Guersent states that spasmodic croup occurs most frequently between the second and seventh years. It differs from true croup in the absence of any considerable amount of febrile disturbance, in the absence of false membrane, and *in the tendency to recur at intervals*, often one or twice in a year. The resemblance to croup quoad the tracheal symptoms in sufficiently remarkable. But there is also dyspnœa occurring in nocturnal paroxysms, as well as the clangey cough and hoarse voice. It is certain that this disease runs in families, and is probable hereditary. It is common in children of highly wrought nervous temperament and in the tubercular diathesis. In such cases the stomach is often found to be at fault, indigestion, sluggish bowels, and depraved appetite being concomitants. The diagnosis of this affection from true croup may be thus tabularly expressed.

Symptoms.	*Croup.*	*Spurious croup.*
Prodromata	Severe	Slight or none.
Fever	Considerable, and without remission	Slight and remitting.
Voice	Faint, whispering, or extinct	Hoarse, never whispering nor extinct.
Expectoration	Mucous and false membrane	None.

Symptoms.	*Croup.*	*Spurious croup.*
Croupal stridulous sound of respiration	Constant	Remits in the interval of the paroxysms.
Dyspnœa	Both in inspiration and expiration.	Chiefly in inspiration.
Recession of chest walls	Considerable	To much less extent.

There is considerable resemblance between this disease and laryngismus stridulus, but they are not identical (as some erroneously teach), for in laryngismus there are no catarrhal symptoms; it is an affection purely spasmodic; whereas in spurious croup there is catarrhal inflammation of the larynx with spasm in addition. Laryngismus, moreover, is a disease essentially of infancy, and often of the rickety diathesis, whereas catarrhal spasmodic laryngitis is an affection common from about the period of the first dentition up till seven, eight, or nine years of age, and affects children of all types, but especially of the tubercular diathesis. Further, spurious croup tends to recur, and this recurrence often leaves its mark on the throat by some thickening of the mucous membrane, causing hoarseness.

Symptoms.—The symptoms of an ordinary attack are as follows:—the child goes to bed with a little feverish cold, perhaps a little hoarseness, and the mother, who has seen several attacks, is on the alert, knowing what to expect. In an hour or two the little one awakes with more or less spasmodic noisy dyspnœa, considerable hoarseness, and a rather harsh cough. The attacks of spasm are sharp, and the child often seems at the time very poorly indeed. It is even sometimes difficult to believe next morning when we find him playing about that it is the little sufferer we saw in the night. But an emetic and warm fomentations are most effectual in this disease. The special aim, however, of treatment must be to prevent recurrence of attacks.

Treatment.—An emetic at the outset, *e.g.*, sulphate of zinc or ipecacuanha, followed by a turpentine stupe or hot fomentations to the throat.

Bromide of potassium combined with an antispasmodic, as lobelia or cannabis indica, or with a sedative, as conium or henbane, may then be given. Nourishing diet and a moderate use of stimulants will be of service. Children who have "repeated attacks of croup" are very susceptible to cold, and very little exposure suffices to induce one of these attacks. Their general health must be improved as much as possible during the intervals, and damp is especially to be avoided.

6. ACUTE SIMPLE LARYNGITIS

is a disease far commoner in adults than in children; still it does occur in children, and requires a brief notice. After rigors and some feverishness, as prodromata, the symptoms are hoarseness, varying from mere soreness up to burning pain referred to the larynx, a sense of constriction, and the child, too young to express itself, pulls at its throat, and gasps for breath; the voice, as the disease progresses, becomes a mere whisper, the child avoids speaking even if it can, as the effort is painful. The breathing is labored, and there is also difficulty in swallowing. Cough when it occurs is spasmodic and violent. The type of fever, which at first runs high, causing hot skin, scanty urine, and quick pulse, soon turns into the asthenic type, from the deficient oxygenization of the blood. The disease ends in convulsions or coma; it may last from four to six days.

Post-mortem.—The mucous membrane is found inflamed and thickened, sometimes abraded, the follicles swelled, the epiglottis stiffened. Sometimes the disease assumes a still severer type, and the submucous tissue of the larynx is œdematous, and the glottis and epiglottis contain serum or pus.

The causes of laryngitis are, exposure to cold and wet, swallowing scalding fluids, extension of inflammation in erysipelas, scarlatina, smallpox, and measles.

Diagnosis from laryngismus stridulus, by the presence of fever and the general symptoms. From croup, by the absence of the peculiar croupy noise and breathing, and by the absence of false membrane.

The prognosis is very unfavorable; the more imminent the suffocation the worse the prognosis; while decrease in dyspnœa, free expectoration, and less difficulty in swallowing are hopeful signs.

Treatment.—A moist warm atmosphere, hot moist sponges to the throat, and tincture of aconite (one drop) every hour or so internally, comprises the treatment I would recommend. I do not think much benefit will be derived from counter-irritation, but there is no harm in using the Liquor Epispasticus over the site of the inflammation. If there be œdema of the glottis (which is common also during chronic laryngitis), the fluid must be let out by scarification, or tracheotomy must be performed. And it should be performed at once without waiting till the child's strength is exhausted. Nutrition by the rectum, when dysphagia renders ordinary nutrition impossible, is to be zealously carried out.

7. CHRONIC LARYNGITIS.

An important and common affection in children is caused by chronic inflammation of the upper portion of the larynx with thickening of the mucous membrane. In this disease the cough is harsh, rough and tearing in character, the voice is rather hoarse. Towards night the cough sounds almost croupy; it is increased by the horizontal position, partly because the uvula is generally relaxed in these cases, and tickles the opening of the windpipe, exciting a cough when the child is put to bed of a tearing, incessant character, often lasting for two or three hours. The subjects of this disease are eminently susceptible to cold; an east wind is sufficient to bring on tearing cough in the night without any coryza or chest symptoms whatever. Sometimes but not always after almost incessant fits of the most violent coughing up comes a small pledget of darkish mucus, with the effect of sudden and complete cessation of the attack. More often it seems to wear itself out, and towards the middle of the night or early morning the child falls asleep, to awake seemingly well in the day but prone at night to a renewal of the disorder.

Flannel next the skin, thicker in winter than in summer but never discontinued, generous diet, and morphia in small doses for the cough itself is the proper treatment of this disease: the morphia even in a few drops often acts like a charm in dispelling the paroxysm, but the tendency to recurrence must be met by appropriate hygienic measures, amongst which flannel next the skin, to prevent chilling of the body, is by far the most important. If the uvula be relaxed an astringent gargle may be used as alum, or it may be touched with solid nitrate of silver, or swabbed over with a solution thereof, or of glycerine of tannin. Sucking ice is very useful in contracting the swollen and congested parts. Epsom salts and tincture of steel should be given internally. I have never got any benefit from expectorants, &c., in this disease. When the uvula remains obstinately elongated in spite of all efforts, the end of it should be snipped off. The annoyance and anxiety of a case like this to both parents and physician until the true cause is discovered and dealt with is great, the severity of the cough being often such that with difficulty we assure ourselves that the lungs are unaffected.

Children are very apt to swallow things they put in their mouths—marbles, coins, &c.; when these, as occasionally happens, "go the wrong way," they usually lodge in the right bronchus. Tracheotomy is the remedy if an emetic and exciting cough will not dislodge them.

8. Bronchitis, or Bronchial Catarrh

is inflammation of the mucous membrane of the bronchial tubes accompanied with increased secretion of mucus. Bronchitis may be acute or chronic, idiopathic or intercurrent.

Acute bronchitis commences with rigors and feverishness, cough, and some pain and tightness in the throat; the cough, at first hard, after a day or two is looser, and in infants especially this materially increases dyspnœa, because they cannot expel the phlegm.

The amount of dyspnœa varies; whether the bronchitis be limited to the large tubes (when it is trifling in amount), or if capillary in character, when it is often most urgent. The pulse is quick, the face is flushed, there is considerable anxiety and restlessness, the countenance may be livid from defective aëration in urgent cases. The disease is very liable to complication with pneumonia; but if it progress favorably, in a few days the fever subsides, the cough is better, and the child progresses to convalescence.

Physical signs.—Percussion is clear, except pneumonia be also present. Auscultation reveals râles, sonorous, mucous, and sibilant; the moist sounds, of course, after the mucous secretion is poured out; large crepitation is heard most frequently at the posterior and inferior portions of the chest. Occasionally a plug of viscid mucus chokes up one of the smaller bronchi, inducing a condition called pulmonary collapse; this accident is manifested by increased dyspnœa *without increased fever*, and the percussion note becomes dull instead of clear, with bronchial respiration. It is important to distinguish this condition from pneumonia, as its treatment is diametrically opposite. In capillary bronchitis, again, the percussion is clear, while on auscultation is heard subcrepitant râle, a moist sound between large and small crepitation in character; with it are often associated both sibilus and rhonchus. Its site is the posterior bases of the lungs, and as the disease advances, it is replaced by large crepitation.

MM. Rilliet and Barthez state that acute bronchitis rarely occurs in children without pneumonia being also present. Hence they say the value of subcrepitant rhonchus as a diagnostic sign of bronchitis differs with the age of the child. If in a child under five this sound is heard on one or both sides of the chest there is danger that the bronchitis is complicated with lobular pneumonia. In older children there is less probability of such being the case. When there is crepitant rhonchus pneumonia is almost certainly present. Bronchitis in children not unfrequently assumes a chronic form with copious perspirations and flushes of fever especially towards night; the disease then bears a strong resemblance to phthisis.

Prognosis.—Bronchitis is dangerous in children at the breast, and under five years of age. Capillary bronchitis, broncho-pneumonia, and collapse of the lung are all very fatal.

Broncho-pneumonia.—Rare during the first year of life; is common after that period up to the fifth or sixth year, when its frequency diminishes. It is a sequela of capillary bronchitis, and also of pertussis, measles, and collapse of the lung. It is more acute in character when supervening on capillary bronchitis; less so when occurring as a sequela of pertussis. The onset—say in the course of capillary bronchitis—is marked by an accession of fever, increased frequency of pulse, and respiration; elevated temperature and orthopnœa. Cough is extremely painful, the face becomes livid, and restlessness and excitability, which are, in fact, a struggle for breath, are soon succeeded by indifference and apathy, passing into coma and death. The disease is very fatal; stimulants, emetics, and stimulating embrocations afford the only hope of success. The physical signs are those of slowly occurring consolidation, dulness, increasing in character, increase in vocal fremitus and fine crepitation, first heard at the bases and spreading over the chest.

Post-mortem.—The mucous membrane of the bronchi is unnaturally red; there is generally some thickening and softening of the mucous membrane; it is covered with thick, grumous, muco-purulent secretion. The tubes are further greatly dilated. There is generally some congestion of the lungs, and traces of lobular pneumonia are common. Portions of lung affected with pulmonary collapse sink in water, are heavy, solid, and deep purple color; on inflation the parts are readily restored to their normal condition. The margins of the lobes are the parts most readily affected by this condition, which is purely mechanical in origin. Lobular pneumonia presents solid, reddish, scattered patches, forming elevations on the surface of the lung. Inflation has no effect on such patches. Frothy or purulent fluid exudes on pressure.

Treatment.—A large sinapism or turpentine stupe, followed by linseed-meal poultices properly made and frequently changed, or a piece of spongio-piline soaked in warm water, are good external applications. Internally in acute cases, a dose of calomel and jalap may be given at the outset, and expectoration should be assisted by ipecacuanha, squill, citrate of potash, senega, and similar remedies; when the secretion becomes abundant, it should be removed from the loaded bronchi by emetics. Mucous and subcrepitant râles are in young children the best indications for emetics, according to Bouchut. Powdered ipecacuanha, mustard, alum, or sulphate of zinc are best; Vin. Ipecac. sometimes fails even in large doses. A warm bath towards evening is good practice, and the child often sleeps after it. If there be much restlessness, small doses of Dover's powder at bedtime are valuable. Should pulmonary collapse take place, stimulants, especially Ammon. Sesquicarb., will be needed, with wine and rubefacient liniments. The diet, at first low, must be improved as the disease progresses, care being taken to eliminate all heavy and indigestible things from what is sanctioned. In chronic bronchitis, where there is less fever, and when the child is already exhausted by illness, the cough will require controlling by such drugs as bromide of potassium, belladonna, morphia, &c. Antispasmodics and anodynes, in fact, in place of expectorants and stimulants. Sinapisms are still useful, and so are stimulating embrocations. Inhalations, too, of steam or medicated vapors (especially that of kreosote) are serviceable. The diet must be light and nourishing, and calculated to restrain rather than promote secretion, and therefore especially limited and defined in the matter of fluids.

In such cases quinine is useful in small doses, or if it disagree, Dr. Meigs recommends the following:

℞ Elix. Cinchon. Flav....................................... ʒij.
    Curaçoa.. ʒij.
    Acidi Sulph. dil.. ℥xij.
    Aquæ... ℥iiss.
Ft. mist., ʒj secund. hor.

In very chronic cases attended with persistent mucous râles over the bases of the lungs, astringents especially gallic and tannic acids are useful. I have certainly seen benefit from their employment. At the same time some external agency should not be neglected, as painting with iodine paint or gentle frictions with some stimulating liniment. Dr. Stierlin of Schaffhausen recommends carbonate of ammonia rather than emetic or other treatment especially in the broncho-pneumonia of young children and in the catarrhal attacks of infants. The dose may be up to 5, 10, or even 20 grains. Dr. Stierlin by this means lost only 7 out of 150 cases, whereas Rilliet and Barthez consider the catarrhal pneumonia of infants especially to be almost invariably fatal.

It is convenient here to mention specially another plan of treatment referred to in former editions as much in vogue in America and on the Continent, but which, since then, has daily gained ground in England and elsewhere, and which is applicable not only to bronchitis, but to most, if not all, acute inflammations, viz., the treatment by the great vascular sedatives, aconite and veratrum viride.

This treatment, according to the evidence of those who have most largely employed it, is most suitable for children over three years of age, whose previous health has been good, and in whom the inflammation is acute and primary. It is a remark of Bouchut's that in the first stage of childhood the material lesions are less purely inflammatory than in the second stage, and the suppuration of the tissues is both less frequent and of a less laudable quality. It is accordingly found that these powerful antiphlogistics are of less service during the earliest years of life. Besides the age of the child, the period of the administration of these drugs is an important point; they should be given *as early as possible* in the course of the disease, in *small* and *frequently repeated* doses, until the activity of the inflammation begins to subside, the pulse lowers, the temperature falls, and moisture appears upon the skin. Beyond this point it is unwise to push the remedy, as the depression so induced may be very considerable. Dr. Lewis Smith, of New York, recommends the following prescription for a child five years old in the first stage of acute bronchitis:

℞ Tinct. Verat. Virid.. ℳxij.
   Syrupi Scillæ comp.. ʒij.
   Syrupi Bals. Tolutan.. ʒxiv.

Misce.—One teaspoonful every two to four hours; the medicine to be omitted, or given at a longer interval, if the frequency of the pulse is reduced.

I have but little experience of green hellebore, but I am more and more satisfied that aconite is a most valuable agent when similarly employed. The dose of the tincture of the Ph. B. may be half a drop to a drop, repeated every hour or half hour until the effect described is manifested. The dose is suitable for a child five years old. When the inflammatory symptoms have abated, the ordinary treatment of the special inflammation and its sequelæ is to be resumed; for example, in bronchitis, expectorant mixtures; in tonsillitis, astringent gargles, and so forth. In fact, aconite and green hellebore thus employed may be regarded as in a measure replacing the depletions and blisterings of days gone by.

9. Pertussis—Hooping-Cough

is a specific contagious form of bronchial irritation and reflex spasm of the air tubes, but especially of the glottis. The usual preliminary symptoms of hooping-cough are—coryza, slight cough, poorliness, pyrexia—sometimes pretty sharp—great nervous excitability, even delirium at night; and these prodromata may last from two to ten days, or may be entirely absent. The cough soon becomes spasmodic, each fit abrupt, and the expiratory efforts may go on to the verge of asphyxia, the veins of the head and neck be swollen, the eyes starting, the nose bleeding, and occasionally the contents of the bladder and rectum are involuntarily discharged. When the spasm relaxes the air rushes in with a full inspiration which gives rise to the sound called the hoop. The paroxysms end in expectoration of ropy gelatinous phlegm like albumen, or often in vomiting, when it is common for the child to ask for something to eat directly, or sometimes in pure exhaustion. In the intervals of the paroxysms the child plays about and appears well. The fits vary in frequency, from one or two in the twenty-four hours, to one or two in the course of an hour, and are brought on either without apparent cause, or by reflex mechanism as in anger, a draught of cold air, the act of swallowing, and such like.

The prognosis is more grave the more paroxysms occur by night. If the case tends to recovery, the cough becomes less spasmodic, the expectoration less glairy, and the disease gradually disappears. It is rare for an attack of hooping-cough to recur in the same individual. It is eminently contagious.

Physical signs of the uncomplicated disease are not very abundant, there is increased dulness on percussion at both bases, and auscultation reveals sonorous and sibilant rhonchi.

Post-mortem.—The disease presents no genuine anatomical character, but there is often found lobular collapse, *i. e.*, the lung is depressed in certain points, in patches from the size of a fourpenny piece to a florin. On section it is found tough and airless, it sinks in water, the grumous fluid of hepatization is wanting, and the lobules may be inflated. Emphysema is not uncommon. Some have described various signs of inflammation about the vagus nerve.

The average duration is about twelve weeks. The disease is generally most severe at the end of the fourth and fifth week. Complications:—bronchitis, pneumonia, croup, convulsions, the exanthemata, especially measles, tubercular meningitis, vomiting, diarrhœa. Death occurs from asphyxia, extreme exhaustion, cerebral congestion, extreme lobular collapse, or capillary bronchitis, or one or other of the complications.

The commoner complications of pertussis require a few words, the very commonest are bronchitis and pneumonia, especially in children of rickety diathesis, and pertussis is then a very fatal disease. There is of course extra fever in such cases, more dyspnœa, especially there is hurried respiration in the *intervals* of the paroxysms, and there is also less vomiting than in ordinary pertussis. Besides these points the physical signs of bronchitis and pneumonia will be present. Dr. Graily Hewitt has shown that pertussis proves fatal not by causing pneumonia but by causing catarrhal inflammation of the bronchi attended with collapse of a portion of the lung tissue.

Convulsions again are common during the course of pertussis, and the violent character of the cough causes also congestion of the brain, and favors any latent tendency to hydrocephalus. Vomiting, when unusually severe and persistent, and especially when accompanied with drowsiness will excite suspicion; if to these be added aversion to light and noise, a dilated pupil, grinding of the teeth, and rolling or tossing of the head, the prognosis will be extremely grave.

Hæmoptysis is not a very grave sign during pertussis, it is not by any means to be looked upon as certainly indicating the presence of tubercles, indeed hæmorrhages from mouth, ears, nose and conjunctivæ are not uncommon. At the same time pertussis occurring in a tubercular or scrofulous child highly favors the production of phthisis and scrofulous affections.

Anasarca is not an uncommon complication also. There is usually some œdematous puffiness under the eyelids, &c., but sometimes this becomes the case in the face and arms and occasionally in the serous cavities fluid is thrown out and also throughout the areolar tissue.

Croup may also develop itself during pertussis, and lastly worms, especially ascarides, often make their appearance in the course of the disease especially in strumous children.

These complications, even the inflammatory ones, as bronchitis and pneumonia, are *not* to be treated by lowering measures. It will be remembered that the rickety or strumous child is commonly the subject of complicated hooping-cough. Sinapisms, stupes, fomentations, with nourishing easily digested food and fair quantities of stimulants, will best meet the chest ailments, while when the complications are cerebral, cold to the head, and the sedative which most effectually quells the cough must be resorted to. Bromide of potassium seems especially indicated under such circumstances.

Diagnosis.—The only affection with which pertussis is at all liable to be confounded is tubercular deposition in the bronchial glands, which indeed may be a sequela of hooping-cough in the tubercular child. Of course bronchial phthisis is not contagious, the hoop is not well marked, and is not followed by vomiting. The disease runs a slower course, and there are the night sweats, evening exacerbations and increasing emaciation of tubercle to aid the distinction. In a case (a girl æt. 12) in which bronchial phthisis supervened in pertussis the hoop remained for many weeks and the violence of the spasmodic attack of cough defied all treatment, was not sensibly diminished by half-drachm doses of liquor morphiæ acetatis frequently given, nor by belladonna, nor by bromide of potassium, nor indeed by any of the remedies resorted to.

Prognosis.—The most danger is in the youngest children, it is more serious when epidemic, and when it seizes upon an unhealthy child.

Treatment.—Probably there is no disease for which so many specifics have been vaunted with such unsatisfactory results. For the early stage (called sometimes the catarrhal stage, and the first stage) the child should by all means be kept to the house; the day and night nursery should be of equal warmth and well ventilated; the child's diet should be light, and if it has been taking stimulants they should be discontinued. For medicine, an ordinary expectorant mixture will suffice. Such means mitigate decidedly the severity of the second or pronounced stage of the disease. For this I am accustomed to rely on the administration of bromide of potassium, or bromide of ammonium, and in cases of especially spasmodic character, I combine one of these with extract of belladonna. Commencing with a grain of either bromide every four hours, it is gradually increased to four or five grains, as the severity of the attack may require. One-twelfth of a grain of extract of belladonna is sufficient at first for a child under twelve months, and it may also be increased cautiously to half grain and even grain doses with great advantage. Acidum Hydrocyan. dil., Succus Conii, alum, nitric acid, valerian, morphia, have all their advocates; and formulæ for the administration of most of them will be found. In my own practice I seldom deviate from the bromides and belladonna, because I am satisfied with the results obtained. Small doses of morphia especially the bimeconate may, however, occasionally replace the belladonna with advantage.

If any particular spasm be very severe and seem to threaten life, cold water should be dashed in the face, and the feet popped into mustard and hot water, or a piece of ice in linen may be applied to the epigastrium according to the plan of Dr. C. D. Meigs. Dr. Churchill recommends half a drachm of sulphuric ether to be spilled in the nurse's hands and held before the child's nose and mouth at the commencement of the fit.

Counter-irritation, or, as some would have us call it, counter-stimulation, is useful in many cases. Roche's embrocation (olive oil one part, and half a part of oil of cloves and oil of amber) is a popular remedy, and an occasional sinapism is of undoubted use. Blisters are an unnecessary method of torture. If the cough be exceedingly severe at night (a grave prognostic), a small quantity of Dover's powder at bedtime is useful. For the latter stages of the disease, when the child is much worn down and harassed by the long struggle, change of air often works like a charm. Cod-liver oil at bedtime is then of service, particularly in delicate children. Complications must be treated as they arise. Pulmonary collapse (known instantly by the sudden general adynamia, urgent dyspnœa, clammy skin, and anxious face, with dulness on percussion) requires wine and carbonate of ammonia, ether or other diffusible stimulants.

Dr. Bottare recommends benzine in pertussis in doses of 20 to 30 drops in mucilage and syrup. It may also, he says, be simultaneously inhaled.

Dr. Bartlett recommends iodide of silver, the dose from ⅛ grain with white sugar for a child 2 or 3 years old, 3 or 4 times a day. Dr. Bartlett in 25 years' experience finds this remedy the most effective and pleasant, and one, moreover, not interfering with the digestive functions.

Hydrate of chloral has been largely used of late in pertussis and other spasmodic diseases; 4 grains given to a child 3¼ years old relieved the sickness of pertussis, but after being given seven successive nights symptoms of poisoning appeared; these, however, yielded to food, stimulants and nux vomica. M. Ferrand recommends 5 grains of hydrate of chloral in simple syrup for a child five or six years old.

Trousseau has pointed out that an acute disease coming on during hooping-cough diminishes its intensity and causes its disappearance either for a short time or in a definite manner. Dr. Ringer speaks very highly of the use of lobelia inflata in hooping-cough, particularly in uncomplicated cases. He gives ten drops of the tincture every hour for a child two years old, and states that the unpleasant effects of lobelia in the adult are not produced in the child. Dr. Ringer further states that the cough is rendered less paroxysmal, and the disease therefore milder and less dangerous. He is more uncertain as to whether its duration is shortened.

10. LARYNGISMUS STRIDULUS—CHILD CROWING

is a disease frequently occurring in the rickety, sometimes in the strumous, child, but often no diathetic disease whatever can be traced.

The invasion of this disease is generally sudden—without warning, and it often comes on during sleep. Sometimes certain prodromata may be observed, as the twitching of the thumb into the palm, a peculiar movement of the muscles of the mouth called "sardonic smile," and slight general facial twitches. Then comes the paroxysm, in which the head is thrown back, the nostrils and mouth are dilated, the veins of the head and neck generally distended, turgid, and swollen; the eyes staring, and convulsive movements occur of the muscles of inspiration; this may last for from a few seconds to nearly three quarters of a minute, during which asphyxia seems imminent, when suddenly the closed glottis relaxes and inspiration takes place with a loud crowing sound, which gives its name to the disease. There is in some cases sudden pallor, as if the child had fainted; without respiratory movements and with but slight lividity—in fact as if the child were suddenly dead; such an attack is more likely to end fatally than the first described. These attacks may end in evidence of cerebral congestion which may be more or less permanent, *e.g.*, hydrocephalus not very rarely supervenes; or they may end in a violent fit of crying, or even sometimes in convulsions, or death may occur from asphyxia or convulsions. The seizures are generally at long intervals, but may be as often as five or six in the day. The mortality is about one in twelve.

This remarkable spasm is undoubtedly of reflex origin and it may arise from irritation of all kinds; as, for example, that of dentition, gastric and intestinal disorders, fright, passion, cold draught of air, the act of deglutition, &c.

Age.—It may occur up to three years, but is rare after twelve months; six to nine months is the common age.

Diagnosis.—From croup. In laryngismus there is no cough, fever, or sign of inflammation, the attack is sudden, the recovery is sudden and perfect; there is no false membrane; croup seldom recurs, while laryngismus often does so. From spurious croup. See that disease.

From Acute Laryngitis.—Laryngitis is rare in infants, is gradual in attack, steady in symptoms, and causes fever and quickened respiration.

Treatment.—The great indication is to remove the cause, as to lance the gums when it is dental; to clear the bowels when it is gastric or intestinal. If prodromata occur the child should be gently aroused and its feet put into warm water, and cold applied to the head. In the actual fit, put the child into a warm bath, and pour a cold dash over the head and shoulders, slap the back, and tickle the fauces to cause vomiting. A piece of ice wrapped in a cloth applied to the epigastrium is sometimes useful, or one of Dr. Chapman's ice-bags may be applied to the spine. In the interval, tonics, especially nervine tonics; cold salt-water baths, regulated diet, and warm clothing. Bromide of potassium undoubtedly checks the frequency of the attacks. Tincture of lobelia and cannabis indica have also been employed in this disease with considerable advantage.

M. Brachet recommends oxide of zinc in half-grain or grain doses with a grain or two of extract of henbane twice or three times a day. Change of air is especially beneficial in checking the tendency to this disease, and when the child is evidently either rickety or scrofulous the appropriate treatment of those conditions will be required to effect a cure. Steel is particularly serviceable, whether as syrup of the iodide or phosphate, in improving the general health and diminishing the tendency to recurrence of laryngismus.

Dr. Gee in a recent article attributes laryngismus to the chilly damp of England, necessitating indoor confinement of children, occasioning as a consequence erythism of the nervous system, and then acting as an excitant on exposure to its influence. In the warm climate of Auckland, where children almost live in the open air, I have seen very few cases of true child crowing, but on the other hand, spasmodic croup is very common. This I believe results from the fact that this atmosphere, though warm, is often very damp.

11. PNEUMONIA

may be primary or secondary in character. There is a variety of Pneumonia in young infants which is not idiopathic, *i. e.*, does not result from any irritation in the pulmonary tissue from atmospheric causes, but results from a stagnation of the blood in the lungs. This is eminently a disease of the poorly nourished, and occurs especially in children lying long on their backs.

When idiopathic, the symptoms first observed are usually restlessness and slight feverishness, which increase in severity towards night, then follow cough, rapid breathing, and great heat of skin, often vomiting, loss of appetite, thirst, and a tongue very dry, red at the tip and edges, and white and furred in the middle. Headache and constipation are common. The hurried breathing prevents the young infant from sucking properly; it takes the nipple for a few minutes, sucking greedily, and then drops it and gasps for breath. It keeps its mouth open to obtain more air, and hence the remarkable dryness of the tongue. Sometimes the access of pneumonia is more sudden; the child awakes in the night with a burning skin, a bounding pulse, a flushed face, and a hacking cough. This form is not generally noticed in children at the breast, but in those a few years older; and when pneumonia so arises it is not unfrequently complicated with pleurisy.

When an attack of pneumonia is ushered in with convulsions followed by loss of consciousness, such pneumonia is usually at the summit of the lung (Rilliet and Barthez). The respiration in such cases is also more gasping and broken than in pneumonia occurring elsewhere. The ordinary number of respirations in the first stage is from 30 to 50 per minute, and the pulse varies from 150 to 160, in the later stages the breathing may be from 50 to 80 per minute. When dyspnœa is very great the nostrils dilate exceedingly, the mouth remains open with its angle drawn downwards and outwards (a very fatal sign) and the face has sometimes a remarkable pallor,—is in fact blanched.

This stage of pneumonia, which is called that of engorgement, then passes into the second stage, that of hepatization. The cough now ceases to be short and hacking, and becomes urgent and painful, the breathing is yet more rapid, and the peculiar working of the nostrils which is presented is an exceedingly characteristic feature of the disease; the skin is moreover very hot (the average temperature being 104° F.) and dry, although the limbs may be cool, while the body is burning, the face looks puffy, and there is a blueness round the mouth, and urgent thirst. Should the disease pass on unchecked into the third stage (that of purulent infiltration) there will be evidences of exhaustion, irregular respiration, cessation of the cough; the face looks sunken, clammy sweats break out, though the skin keeps hot to the last. The pulse becomes so rapid and small that it is difficult or impossible to count it; great jactitation, increased lividity, and either life gradually fails, or convulsions followed by coma put an end to the scene. On the other hand, a fall in temperature on the seventh, ninth, or eleventh day is a favorable sign.

Specially unfavorable symptoms are convulsions, small weak pulse, extreme dyspnœa, persistence of bronchial respiration, excessive and obstinate diarrhœa, and Trousseau adds the occurrence of swelling of the veins of the hands.

The physical signs of pneumonia in the child are, dulness on percussion, especially in the infra-scapular region of the affected side. Dr. West calls attention to a feeling of greater solidity below than above the scapula, and which he says may be perceived before the ear an detect actual dulness on percussion. True pneumonic crepitation is to be heard under the influence of a deep inspiration, but it is not as readily discoverable as in the adult. But subcrepitant râle (a moist sound which is larger than the small crepitation of pneumonia, and smaller than the large crepitation of simple bronchitis) is heard often associated with bronchial breathing. Bronchial respiration rarely disappears before the seventh day, to be replaced by subcrepitant râle. Its persistence beyond that day is an unfavorable sign. If the pneumonia be single (i. e., confined to one lung) there will be loud puerile respiration on the sound side. Should resolution now take place, the bronchial breathing will disappear, the subcrepitant râle become fainter, and be gradually replaced by vesicular murmur. Should the disease pass into the third stage, the bronchial breathing will be found quite to mask the subcrepitant râle, while, if suppuration of the lung occur, large gurgling crepitation may be heard. If the pneumonia be secondary to bronchitis, then we find that from the first there is more dyspnœa, more lividity, more early distress, the cough is more paroxysmal in character, and the respiration becomes irregular sooner than in the idiopathic disease. Subcrepitant râle is heard very extensively over both lungs; fine crepitation is unusual; the disease runs a quicker course, is more severe altogether, and generally more fatal.

Considerable differences still exist amongst medical observers as to the varieties of pneumonia. MM. Legendre and Bailly pointed out in 1844 that many cases hitherto grouped as lobular pneumonia were in reality bronchitis associated with congestion and collapse of the lung, and they described such attacks as catarrhal pneumonia, and instead of dividing genuine pneumonia into lobular and lobar, they describe a lobar and a partial pneumonia. In this they have been followed by many good observers, while others, especially of the French school, oppose this classification. The practical deduction, however, is pretty clear; bronchitis attended with lobular collapse (broncho-pneumonia) is far commoner than either lobar (croupous) or lobular (catarrhal) pneumonia.

The following points embrace the principal practical distinctions which need be noted. Lobar pneumonia is usually confined to one lung and that commonly the right, lobular pneumonia is much more frequent than lobar, and partial than general pneumonia. Lobular pneumonia is usually double, but it may be, and often is, much more extensive on one side than the other. Lobular pneumonia is very much more common in children under five years old. In such it is usually but not invariably preceded by bronchitis. Lobular pneumonia is a very much more serious disease than lobar, which latter, when single and uncomplicated, is rarely fatal, especially when it occurs in children in good health and between the ages of six and fifteen years. In primary lobular pneumonia crepitant or subcrepitant rhonchi are noticeable from the first, and often bronchial respiration on one side towards the base. In the lobular form the subcrepitant rhonchi are more diffused, while bronchial respiration is rarely heard; the latter, however, increases as the disease goes on, while the rhonchi decrease. Bronchial respiration is *per se* a grave sign. In lobular pneumonia subcrepitant rhonchus may be the only stethoscopic symptom present throughout. Lastly, in lobular pneumonia the temperature though high shows more irregularity and more tendency to remit than in lobar pneumonia, and the course is also less definite and more prolonged.

Pneumonia in children is occasionally an insidious disease, its existence is apt to be unsuspected and overlooked. Such is the pneumonia of measles, which affects especially the apex of the inferior lobe, and should be watched and listened for at that spot. Such also is the pneumonia of teething, a disease which, being chronic and accompanied with general wasting, but tumid abdomen, and with little cough at first, is often mistaken for phthisis, or mesenteric disease.

Complications.—Bronchitis may and often does complicate genuine lobar pneumonia, besides those cases in which it is the main and precursory complaint to which lobular pneumonia is secondary. Pleurisy occurs also in about 50 per cent. of the cases and emphysema along the upper portions or free edge of the more inflamed lung. This emphysema is vesicular oftener than interlobular in character.

The *diagnosis* of pneumonia is often a matter of some difficulty from the prevalence of cerebral symptoms, and the absence or comparative slightness of pain and cough.

From bronchitis it is to be diagnosed by the more pungent heat of the skin, the comparative freedom of the breathing—that is, the breathing being less labored in pneumonia though rapid—by the dulness on percussion over the solidified portions of lung, and in the increased vocal fremitus when that sign can be obtained.

From pleurisy. In this disease the vocal fremitus is diminished, and the area of the dulness on percussion often varies with the position of the child; this is never so in pneumonia. Intercostal bulging and displacement of the thoracic and abdominal viscera, together with the diffused bronchial respiration of pleurisy, will aid the diagnosis. Pleurisy is rare under six years, and besides there is in pleurisy absence of rhonchus and the cough is drier.

From broncho-pneumonia, the secondary character, the fact of both lungs being involved, and the fact pointed out by Ziemssen that the whole of the lower lobes are not usually implicated, and that, therefore, dulness is generally confined to the posterior dorsal region, and does not extend so far forward as in lobar pneumonia.

From tubercular meningitis. The pulse is often a valuable guide, being slower in the brain affection than natural, and much quicker in the lung disease. So also the temperature of pneumonia is higher than that of meningitis, and the physical signs will complete the distinction.

From acute tuberculization. The previous history, the more frequent remission in temperature, the more chronic course, and the absence of such marked working of the nostrils as characterizes general pneumonia, will be the main points; but from the coexistence of the conditions diagnosis may be impossible.

Treatment.—The treatment of a disease like pneumonia, which has been made a very battle ground between heroics and expectants, is necessarily a matter of some nicety to discuss. Probably, however, a little common sense may in this, as in so many other hotly debated questions, assist towards a solution of the problem. It should first be remembered (as indeed all admit) that there are pneumonias and pneumonias, differing as widely in their symptoms, import, and gravity, as diseases bearing the same name can possibly differ; from sthenic primary pneumonia to asthenic secondary pneumonia the range is wide indeed, and the treatment will of necessity vary in proportion. It is not often in these days we are called upon to treat the acute pneumonia which our forefathers bled and blistered; when so, we are still taught that the application of a dozen leeches over the chest, followed by a soft warm poultice of bread and linseed meal, is attended with wonderful relief and with the best after consequences. For my own part I can only say that for more than three years in a pretty extensive practice, I have never used leech nor blister to a sick child (and I think three blisters and six leeches represent all I have to answer for in adults during the same period). My recommendation is envelope the chest back and front in linseed-meal poultices. The linseed meal should be the kind with the oil crushed in it, and the poultice should have a little oil smeared over its surface. The poultices should not be too heavy, but moderately thick, well warmed, moist, and changed frequently. When pain is severe I direct the poultice to be made of strong poppy solution instead of water, the poppy decoction to be heated and poured on the linseed meal, just as water is ordinarily used. If some particular spot is very painful, I have a small piece of rag soaked in aconite and belladonna liniments (equal parts) laid next the skin where the pain is, and the poultices applied over. Then internally I administer tincture of aconite in drop doses, very often combined with a little citrate of potash and ipecacuanha wine, and sweetened with glycerine. When perspiration occurs and the pyrexia is manifestly less, I discontinue the aconite. If it be needed, there is no objection to clearing the bowels out at the commencement with two or three grains of calomel, to be followed possibly by a simple lavement. The diet in such a case I keep fairly nourishing, but quite without stimulants, milk, moderately strong beef tea, and plenty of simple drinks like barley water, while thirst lasts.

But if the case be asthenic, say occurring in a child of tubercular, rickety, or scrofulous temperament, I employ a stimulant embrocation over the back, and front of the chest, then I have the whole chest enveloped in cotton wool, and I employ stimulant expectorants, carbonate of ammonia especially, two or three grains every two or three hours, with whatever adjuncts seem indicated. In these cases diet must be far more nutritious, strong beef tea, yolk of egg beaten up with milk, chicken jelly, and stimulants may be required over and above the carbonate of ammonia; if so, good Burgundy is a capital one, perhaps the best, but port wine or a little "three star" Hennessy's brandy will very efficiently supply its place. Never allow sick patients to take the raw common brandy as a stimulant in sickness. It is not much, especially in children, that is wanted, but what little there is, should be *good*.

Pneumonias of "crasis" and "typhoid pneumonias" call for the strongest nourishment that can be assimilated, and for very liberal use of stimulants. It is simply a question of keeping the child alive, and drugs may be forgotten, at any rate until amendment takes place.

I think that convalescence from these acute chest complaints is not sufficiently watched, chill cannot be too rigorously guarded against, the sitting-room and bedroom should be of one temperature, and flannel should certainly be worn next the skin. If medicine is wanted, it will be, above all, cod-liver oil.

12. ATELECTASIS PULMONUM

occurs in infants *congenitally;* the respiration of such infants is feeble, and even intermitting; they wail, but do not cry; they show difficulty in sucking; they are cold and livid, with weak pulse. The disease is simply a portion of lung uninflated; oftenest it is the inferior and posterior portions of the right lung that are so affected. The lung itself is dark red in color, without crepitation, exuding no bubbles, but sanguineous serum, it sinks in water. The cause is defective nerve energy from pressure. The treatment is tappings and frictions before the cord is cut to ensure thorough expansion of the lungs. If, however, the disease be extensive, it often ends fatally. The *acquired* form of this disease is met with oftenest in children about two years old; it is characterized by a hacking cough, dyspnœa, palpitation, epistaxis, melæna, and during inspiration the ribs are drawn inwards and sternum forwards; percussion is dull over the diseased parts. Atelectasis is regarded by many writers as being identical with carnification seen in lungs affected by pneumonia; others consider that it is merely a fœtal condition of the lung remaining for a while undiscovered, and that therefore the disease is never truly acquired. The treatment of such cases demands careful attention to hygiene, especially to ventilation, warmth, clothes, and food. Stimulating expectorants, as ammonia and senega, are useful; so also are stimulating embrocations over the affected surface. It is important that the child should observe a recumbent posture, "to antagonize," says Dr. Rees, "as far as possible the altered movement, and give the best chance for the extended lung again to expand."

13. PLEURISY

as an idiopathic or primary disease is rare during the first five years of life, but it arises secondarily in the course of scarlatina and other acute specific diseases as well as in acute desquamative nephritis, rheumatism and pneumonia, with considerable frequency; moreover, as in the adult, it is occasionally latent. The disease is ushered in with depressions, loss of appetite, and if the child be old enough, rigors; next is noticed acute, sharp pain; the well-known stitch of pleurisy aggravated by inspiration, or coughing, or lying on the affected side and by pressure.

Sometimes vomiting, fever, and a short dry cough are the earliest symptoms; the breathing becomes hurried, the tongue white and loaded, the bowels confined, the pulse hard and quick, the skin hot, the face flushed, and the urine scanty and high colored, the intensity of the febrile symptoms being, however, far less than in pneumonia.

Mr. Crisp considers that a dry rubbing sound with frequent screaming and apparent increase of pain on elevating the head, points very strongly to pleurisy in infants.

Dr. West has pointed out that the pain of pleurisy is often referred to the abdomen, and not to the chest, and that it is attended in such cases with bilious vomiting and purging; this is especially the case when the inflammation is on the right side, and diaphragmatic in origin. The physical signs of pleurisy differ in the child from those usually present in the adult, notably in the absence of friction sound, at any rate during the early stages of the disease. If, however, friction sound becomes audible during the stage of absorption and persists long after the acuter symptoms have passed away, it is a sign that tubercular deposit has probably taken place on the surface of the pleura.

The earliest signs of pleurisy are diminished expansion of the chest, diminution of vocal fremitus, dulness on percussion, and bronchial respiration. MM. Rilliet and Barthez consider bronchial respiration the very earliest audible physical sign.

Bronchophony and ægophony occasionally accompany the pleurisy of children; the latter is ordinarily heard at the lower and hinder portions of the chest, in children above two years of age. When effusion takes place the degree of bulging of the affected side is, by reason of the comparative elasticity of the chest walls, *far greater* in the child than in the adult. The respiratory movements on the affected side are almost abolished, and there is bulging of the intercostal spaces; meanwhile, having more work to do, the healthy lung hypertrophies, so that when absorption takes place the affected side is smaller than the other, and flattened at the infra-clavicular region. This deformity is, however, to a great extent corrected as air permeates the lung more freely.

If pleurisy is about to end in empyema, it is common for a quasi-convalescence to occur after the more acute symptoms have passed off. Suddenly, however, severe dyspnœa sets in, and the child, instead of as hitherto preferring to lie upon the sound side, now lies upon the affected side. The physical signs now show the respiratory murmur to be diminished. If the quantity of fluid is excessive, so as to bind down the lung against the spinal column, there will be no vesicular breathing audible, but bronchial respiration and bronchophony will exist. Sometimes ægophony is heard at the angle of the scapula. If no air enters even the bronchial tubes, *i.e.*, if the lung be completely compressed, no sound will be heard on the affected side, but the percussion will be absolutely dull, and the motions of the chest will be notably diminished. On the healthy side the respirations will be intensely puerile. It is common for the matter to find an exit through the chest walls, though occasionally an empyema has emptied itself into a bronchus, or has even burst through the diaphragm into the peritoneal cavity, causing, of course, fatal peritonitis. Generally, however, the matter is let out through the anterior wall of the chest, very commonly between the fourth and fifth ribs, a little outside the nipple. The opening, whether natural or artificial, is apt to remain fistulous for some time, and to produce by the sudden contraction of the side considerable and even permanent deformity.

The *post-mortem* appearances are similar to those found in the adult; the pleura is found smooth, pale, and semi-transparent, or sometimes finely injected; adhesions between the pleuræ; effusion of serum, sometimes transparent, often reddish, sometimes sero-purulent fluid with flakes of lymph floating in it; the costal and pulmonary surfaces both coated with lymph. Often the anatomical characters of pneumonia are superadded, more rarely pus is found.

Diagnosis.—The cases in which the origin of the disease appears abdominal will be best recognized by the dyspnœa, by the cough, and by careful attention to the physical signs. Sometimes in young children more especially, the attacks appear cerebral; here again the dyspnœa and the physical signs will correct the diagnosis.

From pneumonia; something must be gathered from the aspect of the child, its breathing labored, short, and quick, but without the working of the nares so remarkable in pneumonia; the attitude is that of a child dreading to breathe because of the pain, and even holding its chest; but it must be remembered that pleuro-pneumonia is by no means an uncommon condition; on the contrary, that the secondary pneumonia of measles, for instance, is often complicated with pleurisy. A good guide is that given by Dr. West:—"A case is pleurisy which presents sudden and severe symptoms of pneumonia, but in which auscultation fails to detect fine crepitation, and discovers only feeble respiration on one side, and bronchial respiration on the other."

In lung consolidation from pneumonia the breathing is harsher and more tubular, vocal fremitus and vocal resonance are increased. Respiration again may be half as frequent as the pulse in pneumonia, seldom more than one-third as frequent in pleurisy, and the temperature is higher in pneumonia. The line of dulness also varies with the position of the child in pleurisy and does not do so in pneumonia. Ægophony is characteristic of pleurisy.

Lastly from hydrothorax, by the absence of pain in the latter, by the effusion being usually double, and by the fact that hydrothorax is almost always secondary to some acute disease, as nephritis, &c., and attended with effusion of serum elsewhere.

Prognosis.—Intercurrent pleurisy is more serious than the idiopathic disease, and pleuro-pneumonia is the most serious of all. The younger the child generally the greater the danger. Death sometimes occurs suddenly in cases of considerable hydrothorax, especially when the serous effusion has been rapid.

Treatment.—As in other acute diseases. I am not prepared to say that the use of cupping or leeches will not give the relief accredited to them. I can only say that even in the healthy young adult I have used no bleeding, and but few blisters, though some acute cases have fallen to my share. Dr. Roberts advocates the fixing of the side affected by strips of plaster four or five inches wide. They should be long enough to reach from midspine to midsternum. The patient should breathe out, and beginning below, a strip is fixed parallel with the ribs, then one over this, but across the course of the ribs, the third is as the first, overlapping half its width, the fourth follows the second and so on. This is the treatment by "rest," which is also advocated in phthisis and other chest affections. While the pain is severe I employ an aconite and opiate lotion on a piece of rag, applied where the pain is, and over it linseed-meal poultices, or turpentine stupes may be employed, or a blister made by blistering fluid if thought specially desirable. I can remember one case in which nothing seemed to give relief till a blister was raised, and some recurrence of pain taking place later on in the attack, the young lady *asked* to have another blister, though the first was scarcely healed. I am satisfied also of the relief which calomel and opium often give in acute pleurisy, and in a specially acute case I see no objection to their employment. A small dose of calomel and a little Dover's powder is the best form for children. But as in other acute inflammations, I think aconite of great value, and I think in pleurisy I can speak well also of Veratrum viride, but the Veratrum especially must be discontinued so soon as decline in pyrexia is established. Iodide of potassium in combination with a diuretic, *e.g.*, the citrate of potash, then becomes useful, and digitalis, especially the *fresh* infusion, may be employed with benefit. Dr. Anstie recommended tincture of steel. When effusiom has taken place the best methods to procure absorption are the use of a succession of flying blisters, or painting with iodine paint. Should, however, these means prove useless, the symptoms continuing urgent, the child becoming weaker, and the lung, evidently seriously compressed, giving rise to great dyspnœa, the question of paracentesis thoracis arises; and much as has been said on both sides of this important question, the balance of recent opinion is undoubtedly in *its favor*, when carefully performed in properly selected cases, and not left until the powers of nature are exhausted in seeking to make a natural outlet for the escape of the pus.

The late Dr. Anstie summarizes the matter in the following propositions in his article in Reynolds' "System of Medicine:"

1. In all cases where the fluid is so copious as to fill one pleura and begins to compress the lung of the other side.

2. In all cases of double pleurisy when the total fluid probably occupies a space equal to half the dimensions of the two pleural cavities.

3. In all cases where the effusion being large there have been one or two fits of orthopnœa.

4. In all cases where the contained fluid is suspected to be pus, an exploratory puncture should be made and the fluid let out.

5. In all cases where a large effusion has existed a considerable time (say a month) and there is no sign of progressive absorption. (This is rare in children.)

It is always advisable when the operation has been determined upon to make a preliminary puncture with a grooved needle; a small trocar and canula may afterwards be used. The place usually recommended, supposing no reasons exist against it, such as adhesion, &c., at the spot, is the intercostal space between the fifth and sixth ribs just behind their angles; the puncture may, however, be made elsewhere, as, for example, posteriorly between the ninth and tenth ribs. An incision about one inch long is made through the skin and muscle, and the trocar plunged into the pleura; this should be done near the upper margin of the rib because of the intercostal vessels. If a drainage tube be employed to empty the pus as it forms, an eyed probe is passed through the aperture, and made to project at one of the intercostal spaces, as far back and as low down as possible; here it is felt for, and cut upon; and then a piece of silk, attached to the eye and drawn through the wound, and by means of the silk the india-rubber tube, the ends of which are tied together and the pus allowed to drain. The result of the drainage tube, however, is by no means uniformly satisfactory, it is apt to render the discharge profuse and exhausting by continual irritation.

A better plan is to employ Bowditch's exhausting syringe; the cavity should be evacuated without the admission of air, and closed. If the pus be foetid a counter-opening may be made and a drainage tube introduced; injections of iodine have the effect of diminishing the foetor, it is doubtful if they exert any influence in closing the cavity; one part of Tinct. Iodi to seven of water is a good strength to begin with, gradually increased to one to four, it often averts the secretion of pus and no pain follows its use. During convalescence, cod-liver oil will be of service to prevent the deposition of tubercular deposit and to aid in restoring the nutrition of the child.

14. Phthisis.

The following are the noticeable points of difference between the deposit of tubercle in the young and in the adult as summed up by Dr. West.

1. There is a marked difference in the liability of various organs to the affection in the adult and in the child.

2. Tubercle is usually deposited in a greater number of organs at the same time in the young than in the adult.

The lungs and, in about a quarter of the cases, the bronchial glands, are the organs most usually affected in adult life, and the deposit is most frequently confined to them.

Then the tubercle of childhood differs in kind from the tubercle of the adult.

1. Gray granulations and crude miliary tubercles constantly exist by themselves in the lungs of a child; this is rare in the adult.

2. Yellow infiltration of tubercles, "cheesy tubercle," is very commonly found in childhood, and this is often limited to one lobe, generally the upper; it is, moreover, commonly associated with advanced tubercularization of the bronchial glands.

3. Cavities are rarer in children than in adults.

4. Tubercle in the bronchial glands is not only commoner than in the adult, but it occurs occasionally as a primary affection in the child, so as to outstrip in importance the deposit in the lungs.

Symptoms.—The disease commences insidiously; the child "droops," to use a very expressive word of mothers, is languid, irritable, and complains of pain all over the body. A little cough comes on, there is no expectoration, because children swallow their spittle. There is no hæmoptysis, there is seldom diarrhœa, and only occasionally profuse sweats. There is, however, more dyspnœa than in the adult, more general feverishness, more rapid wasting.

Dr. Ringer points out that a persistently high temperature for weeks together, say of 103° F. or upwards, night after night, is a strong presumptive evidence of tubercular deposition.

The skin soon becomes wrinkled and the face old-looking; intercurrent bronchitis or pneumonia is common.

The pneumonia which often complicates phthisis is to be diagnosed from simple pneumonia; in the complicated disease the temperature is lower, the heat of the skin less pungent than in the simple disease, and moreover, the dyspnœa is out of proportion to the amount and gravity of the pneumonic signs. The pulse is also less frequent, and finally there is the history and type of the tuberculosis to complete the differences.

When the bronchial glands are much affected, the attack is marked by more irritative and spasmodic cough, more catarrh, more dyspnœa, and greater general suffering. Hæmorrhage may occur from the suppuration of a bronchial gland involving a vessel. Very frequently evidence is afforded of tubercularization elsewhere, as in the peritoneum or the brain.

Physical Signs.—Inasmuch as the apices of the lungs are not so frequently specially affected in the child, but the deposit of tubercle is more diffused throughout the lung substance, the physical signs vary in the child from those relied on in the adult. There is usually general, though slight dulness on percussion; some flattening under the clavicles; the expiratory murmur is prolonged, and the breathing interrupted. Vocal fremitus is a sign of less value and reliability in the child than in the adult. When present it affords strong evidence of solidification, but it is frequently absent, and yet solidification may exist. At the commencement of the deposition of the tubercles the breath sound is weak, or bronchial—often with a click at the close of the act of inspiration. Later, moist sounds of various kinds are audible, râles, sibilant, mucous, and subcrepitant, over one or both sides of the chest; and later still, as the tissue breaks up and cavities form—mucous râle, cavernous respiration, gurgling, and occasionally pectoriloquy. Bronchial breathing is a sign of especial significance when heard away from the interscapular region where it is a normal condition, as at the apex or base. MM. Rilliet and Barthez consider that harsh and prolonged respiration with increase in vocal resonance are the most significant symptoms of crude tubercle, particularly when heard over the greater part of the lung, or at any rate not confined to the apex. If harsh breathing persist for several weeks and is then succeeded by weak or interrupted respiration, or that form of inspiration which terminates with a click, the evidence of tubercular deposit is extremely strong.

Bronchial phthisis is especially indicated by dulness on percussion between the scapulæ; if this coexist with resonance and fairly good respiration over the upper portions of the chest, and prominence of the veins of one side of the neck, in particular a full jugular during the act of coughing, the probability of bronchial phthisis is extremely strong. The venous engorgement is due to the pressure of the enlarged glands upon the innominate or superior vena cava. At the same time bronchial and pulmonary phthisis often coexist, and the physical signs may become subject to great variety, and the diagnosis be a matter of much difficulty.

Bronchial phthisis is commonest between the ages of two and six years. Special symptoms may be caused by pressure of the enlarged glands in different directions, *e.g.*, œdema of the lungs, hæmoptysis, and even hydrothorax from compression of the vena azygos (Jenner).

The prognosis is always serious, but seldom hopeless; if the disease runs a course unusually rapid in the young, so, on the other hand, medicines are more powerful and the reparative processes more vigorous than in adult ages.

Causes.—Predisposing causes are:—hereditary tendency, cold, damp, deficient food and clothes, vitiated air, and generally bad hygienic conditions. The exciting cause is often an attack of bronchitis or pneumonia, or some eruptive fever. Acute phthisis is especially apt to follow lobular (catarrhal) pneumonia. The average duration of the disease is stated at from three to seven months in children.

Treatment.—When hereditary tendency exists the greatest care and attention must be paid to all the minutiæ of food, nursing, &c., which makes so much of the difference between healthy and ailing childhood; cold and damp are to be avoided, flannel should be worn next the skin. Contagious disorders, especially pertussis, are greatly to be dreaded and studiously to be avoided; climate must also be studied. The greatest benefit often attends the sending of tubercular children away for the winter months to Nice, Pau, Mentone, Barcelona, Hyères, and other spots, nice judgment being needed in recommending the climate best suited to the individual case. In this country, Ventnor, Torquay, Penzance, and Bournemouth, with perhaps Hastings, are desirable winter residences when foreign climates are debarred. Again, all digestive disorders in tubercular children require the greatest attention and should be corrected without delay; no diarrhœa should ever be permitted to go on unchecked. As to actual drugs there are five or six on which much reliance is to be placed; foremost amongst them, of course, cod-liver oil. When this disagrees, as it occasionally will, give it how we please, glycerine or cream may be substituted. The syrup of iodide of iron and of the phosphate of iron are very valuable remedies. Quinine also and bark, in many cases of defective appetite and general nervous debility and depression, are most useful. Iceland moss may be tried as a dietetic. The cough must be checked by anodynes. Small doses of morphia, belladonna, and hydrocyanic acid are the best remedies, with expectorants when needed, for occasional bronchitic attacks. If there be night sweats, the mineral acids should be given with bark. If diarrhœa, the mineral acids or an enema of starch and opium. Painting under the clavicles or on other portions of the chest with iodine is often useful; and the same may be done with stronger paint between the shoulder-blades behind, when there is the irritative cough of bronchial phthisis. Salt-water bathing, warm or cold according to the weather and the strength of the patient, is very beneficial. The food must be nourishing yet unstimulating; milk, butter, and fats when willingly taken are useful, but they should not be forced. Local pains may be relieved with occasional sinapisms; blisters are better avoided. Stimulating embrocations are of good service, and it is often useful to rub the body with cod-liver oil when that medicine cannot be tolerated by the stomach. In such cases cocoa-nut oil has been recommended as a substitute; it may be taken in doses of one or two teaspoonfuls twice a day. The hypophosphites of lime, soda, manganese, and iron have been much employed by Dr. Churchill, of Paris, and other Continental physicians. Pepsine and pancreatic juice are of service in the dyspepsia so often incidental to phthisis, and which is so serious an impediment to the due nourishment of the patient, a point, indeed, of vital importance.

Dr. Henri Blanc administers the juice of raw meat, alcohol, and phosphate of lime. The following is the process he recommends for preparing the raw meat. A pound to a pound and a half of fresh beef, deprived of fat, bones, &c., is placed over a quick fire for a few minutes, in order to whiten and harden the external surface only; the piece of meat is then cut into two or three pieces corresponding to the size of the meat press, and all the juice is extracted by the pressure of the powerful screw. The superficial coction is necessary to overcome the elasticity, which renders the extraction of the juice a very difficult matter unless more powerful machines be used than the simple one at present required. A pound and a half of good fresh meat gives a teacupful of juice. The juice should be prepared daily. This juice, having all the physical properties of raw meat, is easily digested, is well tolerated, and, served in the following manner, is always very grateful to the patient. The juice should be mixed with equal parts of tepid broth, made of bones, and flavored with salt and pepper, and to which tapioca, vermicelli, &c., can be added. Care, however, should be taken that the broth is never more than *tepid*, otherwise coagulation takes place, and the desired effect is not obtained.

Modifying Dr. Blanc's treatment to the needs of children I should say it would stand thus:—Early morning, warm milk with bread and butter, and, if fancied, an egg. At 8 or 9, breakfast (before which a half drachm of Easton's Syrup of the Triple Phosphates is to be taken); during this meal some phosphate of lime is to be taken, and half the daily allowance of raw meat juice in beef tea or broth. The breakfast itself should consist of a good nourishing meal—fish or poultry, fresh vegetables, and a little wine, I would say a good sound claret, or the Greek santorin, or a little real manzanilla. Dinner at about 2 o'clock, also a good nourishing meal, and soup or broth with the remainder of the raw meat juice, and, instead of the triple phosphates, a dose of cod-liver oil may be tried about an hour after it. A little phosphate of lime is to be got down during the dinner; it is almost tasteless between thin slices of bread and butter. The last meal, or tea, is to be again warm milk fresh from the cow, and diluted, if necessary, with a little Vichy, Friederichshall, Apollinaris, or other gentle aperient. I must say that I have come across few more rational or hopeful methods of treating advanced phthisis. At least the physician will have the satisfaction of feeling that he did his utmost to sustain and to prolong life.

15. CYANOSIS, OR BLUE DISEASE

is a bluish discoloration of the skin arising in connection with malformation of the heart, such malformation being usually due to permanence of the foramen ovale, or to abnormal apertures in some part of the auriculoventricular septa, or to the origin of the aorta and pulmonary artery from a single ventricle, or to permanence of the ductus arteriosus. When such patients survive they suffer from coldness of the body, palpitations, dyspnœa, syncope, congestions, and dropsies. Of 186 cases collected by Dr. Lewis Smith, 67, or more than one-third, died before the close of the first year; 121, or more than three-fifths, before the age of ten; only 24 survived the age of twenty, and 4 the age of forty years.

The mode of death in the majority of cases appears to be a sudden paroxysm of dyspnœa. Convulsions, especially in infants, hæmorrhage, and coma, are also common terminations. It is remarkable, as disproving any antagonism between cyanosis and heart disease generally and tuberculosis (in which antagonism even Rokitansky has declared himself a believer), that in 13 per cent. of Dr. Lewis Smith's cases, tuberculosis was present, and in several the lungs actually contained cavities.

Treatment must be, of course, palliative, nourishing diet, warmth of climate and of clothing; avoidance of fatigue and excitement, in addition to which something may be hoped for from posture; that in which dyspnœa is most avoided being adhered to, or instantly resorted to when paroxysms threaten.

16. PERICARDITIS, CARDITIS, AND ENDOCARDITIS.

These diseases are not common in children because rheumatism is not common, nor is renal disease common, nor are the many conditions that offer a mechanical impediment to the onward flow of the blood in mature years common in children. Yet pericarditis intercurrent in rheumatism, in scarlatina (and then probably always connected with slight rheumatism), in measles, in diphtheria and disease of the kidneys, is of sufficiently frequent occurrence to require careful consideration. Under the head of rheumatism will be found an account of the more insidious form of pericarditis occurring with but little pain or uneasiness; but sometimes the disease is manifested by severe pain and much fever, pain in the heart, pain shooting to the shoulders, pains shooting down the arms. Palpitation also occurs, and the heart beats irregularly and with heaving impulse; the breathing is rapid, the face anxious, the head aches, the temples throb, and there may be syncope, paroxysms of impending suffocation, bleeding at the nose, or hæmoptysis.

At the same time it must ever be borne in mind that the slightest forms of rheumatism, or perhaps to speak more correctly the slightest manifestation of the rheumatic diathesis—*e.g.*, Torticollis or Erythema nodosum—may be complicated with heart disease. Other causes of peri- and endocarditis are an inflammatory action spreading from the lung and pleura. Pyæmia rare, pertussis and rickets lead to undue pressure on the right heart cavities, and produce dilatation and hypertrophy.

The physical signs will vary with the portion of membrane or heart substance attacked; in pericarditis there are sensations of friction perceptible to the touch—a rubbing sound, "to-and-fro" sound—preternatural dulness over an enlarged area resulting from the effusion of serum. If the endocardium be attacked, "murmurs" will be audible, and these will indicate diverse conditions according to their site. The following table will sufficiently indicate the differences between the more common forms of valvular disease:

Base or Aortic Murmurs.

If Systolic = Constrictive disease } Pulse, regular, jerking, visible.
If Diastolic = Regurgitant disease }

Apex or Mitral.

If Systolic = Regurgitant } Pulse, irregular, soft intermittent, with pecu-
If Diastolic = Constrictive } liar thrill.

Mitral affections are far commoner in children than aortic, especially mitral regurgitant disease. Aortic regurgitant is rare and seldom exists alone. Mitral regurgitant and obstruction together are not uncommon. In children affected with mitral disease the præcordia is very prominent. Hypertrophy is rare alone, it is commoner with dilatation, to which indeed children are especially liable. Croup, laryngitis, broncho-pneumonia, capillary bronchitis, pertussis, &c., all tend to cause dilatation of the right side of the heart. Palpitation in the child never gives rise to the angina of the adult, and dropsy is later in its advent. Reduplication is considered pathognomonic of adherent pericardium.

It is important to distinguish the dulness of pericardial effusion from the dulness of hypertrophy of the heart; this may be done by noting that in

Effusion.	*Hypertrophy.*
The dulness extends upwards to the second rib, but downwards not much below the natural limit, and changes day by day. There is, moreover, thrill always present.	This dulness extends in all directions, and is stationary, and the disease is characterized by a peculiar heaving impulse.

If the substance of the heart be affected the beat is tumultuous and irregular, and fatal syncope often occurs. Carditis rarely occurs alone, but is usually associated with more or less peri- or endo-carditis, or both. The dulness of endocarditis, which is rare, is distinguished from that of pericarditis, in that the beat appears superficial, while in pericarditis it is remote.

The result of these inflammations is of course more or less damage to the structures of the heart, permanent valvular disease, induration from lymph deposits, saccular dilatation of the heart-walls, rarely abscess, very rarely rupture. General dropsy, cerebral affections, &c., are also among the sequelæ.

Prognosis.—Unfortunately, as a rule,[1] the condition of the injured valves and textures goes on from bad to worse, and the child dies after a few years of increasing suffering. But occasionally the progress of the disease is much slower, or even apparently stationary; and Dr. Latham suggests "that there may be a certain protective power inherent in the growing heart to accommodate its form and manner of increase to material accidents, and so repress or counteract their evil tendencies."

It would indeed appear that the presence of hypertrophy and dilatation of the heart, or of dilatation without hypertrophy, materially affects the prognosis, rendering it more grave, and impressing the necessity for that absolute condition of rest which appears most unfavorable to the production of a dilated condition of the heart. And this is of the more importance, inasmuch as dilatation is more readily caused in the heart of the child than of the adult by reason of its greater feebleness, its more rapid circulation, and its proneness to palpitation and excitability.

It will be well, before considering the general treatment of diseases of the heart, to summarize the conditions that may occur, and their diagnosis.

1. Simple hypertrophy, *i.e.*, the thickening of one or more of the chambers of the heart without increase in the size of the cavities.

2. Hypertrophy with dilatation, *i.e.*, the walls thickened and the cavities enlarged.

3. Concentric hypertrophy, which is congenital, and in which the walls are thickened and the size of the cavities diminished.

The left ventricle is the cavity most commonly hypertrophied. Hypertrophy and dilatation of the right ventricle is usually due to some lung disease, offering mechanical impediment to the onward flow of the blood.

Dilatation is also of three kinds:

1. Active, in which the dilatation is in excess of the hypertrophy.
2. Simple, where the walls of the heart are of natural size.
3. Passive, in which the walls are thinned.

Among the most valuable physical signs of these conditions are two, viz., heaving impulse as indicative of hypertrophy, and increased dulness on percussion as indicative of dilatation. With these will be combined more or less dyspnœa, giddiness, and tumultuous beating of the heart, but in hypertrophy there is increased impulse, and the apex-beat prominently visible, whereas, in dilatation, especially passive dilatation, the impulse is feeble and the apex-beat scarcely to be seen, and, moreover, general dropsy soon supervenes.

Treatment.—The treatment of affections of the heart, acute and chronic, is a very large subject for a very small space. To commence with acute affections, that of rheumatic pericarditis has been already spoken of. When the disease is acute and idiopathic, we are generally advised to place a few leeches over the pericardium and to follow this up by using blistering fluid, and the administration of gentle or smart purgatives, iodide of potassium and diuretics; calomel and opium in small doses are also largely recommended and I have seen benefit from their use. I think, however, that those who abandon this routine treatment, and feel their way carefully but steadfastly with that about to be discussed, will not regret doing so. First, then, locally, maintain constant warmth and moisture by the use of linseed meal poultices, made, if necessary, with poppy decoction, as before described (pp. 231–2); have these frequently changed. See that everything about the little patient conduces to mental and bodily repose. Keep the diet light yet nourishing, and by no means withhold wine when the general state seems to demand it. While the inflammation runs high, we have two great remedies, Veratrum viride and aconite. Valuable as aconite indisputably is, given as elsewhere recommended, in small doses often repeated, I am inclined to believe veratrum even more so in this affection. Small doses, half a drop or less, of Keith's concentrated tincture every two hours exerts a decided and beneficial influence. So long ago as 1868 Dr. Waring-Curran states in the "Practitioner" that he had been led to regard it almost as a *specific* in pericarditis. It reduces the frequency of the pulse and increases the renal and hepatic secretions. But the effects of so powerful a remedy must be narrowly watched, and any signs of failure in the heart's action, too rapid falling of the pulse, &c., may be met by changing the remedy to digitalis, more particularly if symptoms of dropsy should manifest themselves. Digitalis is a cardiac stimulant in one phase of its many-sided and at present incompletely understood actions on the heart. One to five drops of the tincture every four or six hours (better, a corresponding dose of the *fresh* infusion) in children from one to five years old (a drop for a year), will control a fluttering, inordinate action of the heart. The pulse becomes stronger, more regular, and slower, while the urine is largely increased. Belladonna extract over the heart will assist to control excited action when necessary. Any signs of obstruction from coagulation (manifested by orthopnœa, anxiety, and tendency to syncope) are to be met with carbonate of ammonia and stimulants freely given.

For the removal of effusion, flying blisters are commonly recommended, or strong iodine paint, with iodide of potassium, and diuretics, especially digitalis. Tapping the pericardium and injecting with tincture of iodine has been successfuly accomplished, and might possibly be resorted to in cases of desperate effusion.

In the many secondary conditions that may arise, such as congestion of the lungs, pleurisy, hemorrhage, congestion of the liver, &c., special means, impossible to be detailed here, will be required, and will vary with each case; with regard to hæmorrhage the value of digitalis in controlling it, must not be lost sight of. Ice, too, so grateful in almost all illnesses, will be a valuable adjunct. With regard to hypertrophy and dilatation, their occurrence may be frequently avoided by enforcing absolute *rest*. When present, Dr. H. C. Wood, of Philadelphia, says, "Nothing is more marvellous in clinical medicine than the relièf that may be rapidly afforded in cases of simple dilatation of the heart by digitalis." The flagging circulation improves, the rapid irregular beat becomes steadier, the lungs become clearer, the dyspnœa abates. Rapid pulse and weak heart call for digitalis, and cardiac general venous congestions and dropsies call for digitalis. On the other hand hypertrophy requires rather veratrum or aconite. Da Costa says "in much irritability with slight hypertrophy a combination of aconite and digitalis did good." On the other hand, there is no doubt that many cases of hypertrophy are benefited by digitalis. Irregularity is another heart symptom calling for digitalis, but irregularity is not very common in young children. It would be entering too wide a field to attempt to speak in detail of the many and varied conditions of valvular disease. As a matter of fact digitalis will afford relief in a very large number of such cases, as I have repeatedly tested.

Take the following as a typical case admirably sketched by Dr. Ringer: —"A child, suffering from mitral regurgitant and obstructive disease, with ventricular dilatation, but chiefly in the right side. There is unintermitting inability to lie down, paroxysms of severe palpitation lasting hours or days, with a pulse at these times very frequent, but always regular. Dropsy sets in; digitalis checks the palpitation and removes completely the extensive dropsy by greatly augmenting the kidney secretion, and gives great relief. In this condition, sometimes better, sometimes worse, the patient may remain for years, and then the pulse becomes irregular, but without any increase in the severity of the symptoms. These from time to time recur, but on each occasion gives way promptly to digitalis as soon as it has greatly lessened the frequency of the pulse."

Such a sketch pondered over is singularly instructive. Dr. Ringer elsewhere draws the conclusion, "It is impossible to foretell how much benefit digitalis will confer in cardiac affections, or how long the benefit will last." In other words, in the present state of knowledge, its use is tentative and should be carefully watched. The smaller the dose commenced with the better, both for caution's sake and even more because a long use of the drug may lie before the patient, and to produce relief augmentation may, as time goes by, be required. It is needless to add that all hygienic matters should receive special attention in the subjects of heart disease—good nourishment, fresh air without undue fatigue, a little sound Burgundy or port, when needed. The child's attention is not to be called to its heart, as some foolish mothers do, but it is to be made to understand that it is not a strong child and must therefore not romp and play with the rest. A wise mother will compensate, unobtrusively, to the invalid much of what its affliction debars it of. Belladonna is useful in the irritative cough and troublesome neuralgic pains such little ones often experience, and henbane and bromide of potassium are the best remedies for sleeplessness and nervous excitability frequently observable in these delicate and susceptible children. Lastly, hope is not readily to be lost; the conservative forces of nature are powerful, and many cases of chronic valvular disease attain maturity, and last, like the proverbial cracked pane of glass, as long or longer than the sound ones.

17. Epistaxis

may be primary or secondary.

When primary, is never dangerous.

As a secondary condition it occurs in purpura in the course of the acute specific diseases, especially typhoid; in pertussis; in valvular disease of the heart, &c.; it is then a more serious affection.

Should treatment be necessary, ice to the forehead and spine, syringing the nostrils with iced water or a decoction of matico, or if obstinate, plugging the nostrils with pledgets of lint soaked in perchloride of iron, and pushed back as far towards the pharynx as possible. If this be effectively done, it will seldom be necessary to plug the posterior nares, which, however, may be resorted to as a last resource.

It should, of course, be remembered in secondary epistaxis that, besides local treatment, constitutional remedies will be required, directed to the special condition which may exist.

CHAPTER VIII.

DISEASES OF THE FOOD-PASSAGES AND ABDOMINAL ORGANS.

1. THRUSH

is an affection very common in young infants, more especially in those brought up by hand. It is chiefly of importance as evidencing impaired nutrition. The mucous membrane of the mouth is covered with numerous white specks like small atoms of curd; these are most abundant on the inner surface of the cheeks and on the tongue and fauces. They get larger for a few days, fall off, and are rapidly reproduced. The infant's mouth becomes hot, the lips swollen; there is dribbling of the saliva; moreover, there is generally coincident some gastro-intestinal disorder, often green evacuation. The acridity of these motions causes an erythematous blush around the anus, and it is not uncommon to find aphthous spots upon the edge of the mucous membrane of the bowel. The thrush is then said to have "passed through" the child.

As in adults, an appearance of a crop of thrush is often but one indication amongst many of the general asthenic condition of the child; it is therefore in such cases secondary in character.

Professor Berg, of Stockholm, was the first to discover the *Leptothrix buccalis* and the *Oidium albicans*, cryptogamic growths always present in the white specks. The growth of these plants appears to be favored by disturbance of digestion, subacute inflammation of the mucous membrane of the mouth, and acid secretions. As the buccal secretions of the infant for the first six weeks are acid, the prevalence of thrush during that period is thus explained.

Treatment will comprise attention to the general constitutional condition. The bowels are to be regulated, acidity of stomach corrected, great cleanliness observed in all vessels and articles used for feeding the infant, as spoons, bottles, &c. The mouth should be gently cleansed after each feeding with a piece of soft rag. Locally, a solution of borax ($\mathfrak{Z}$ ss.— $\mathfrak{Z}$ j.) is to be applied with a camel's hair brush two or three times a day. This is better than Mel Boracis, as it is a question whether the honey undergoing fermentation does not increase the mischief. For very young infants a grain or two of borax with a little loaf sugar may be put on the tongue, where it dissolves and is swallowed. This plan is easy of management and very effectual; the borax may also be used with glycerine, and the mother should bathe the nipples after suckling and anoint them with glycerine of borax. We are indebted to Sir W. Jenner for another and very efficacious remedy in sulphite of soda; the secretion of the mouth being acid, the sulphite is decomposed and sulphurous acid set free, which destroys cryptogamic plants. The strength of the solution may be $\mathfrak{Z}$ j.

of the sulphite to ℥j. of water. Guersent recommends in severe cases a lotion of—

℞ Sodæ Chlorinat... ℥j.
    Decocti Cinchonæ... ℥iij.
    Syrupi Cort. Aurant....................................... ℥j.
Fiat lotio.

Cauterization with nitrate of silver may be resorted to in extreme cases, and when very obstinate, change of air is eminently beneficial.

2. STOMATITIS, OR INFLAMMATION OF THE MOUTH

occurs in three varieties, as it affects respectively the mucous follicles of the mouth, the substance of the gums, or the cheeks.

1. *Follicular or Aphthous Stomatitis* occurs idiopathically from a heated and disordered stomach, or as a sequela of one of the eruptive fevers, especially measles. It is rare after five years of age. The disease manifests itself by a difficulty in sucking and swallowing, increased flow of saliva, and by tenderness in the region of the submaxillary glands. With these symptoms are also feverishness and restlessness, and often feculent diarrhœa. Inside the mouth and on the tongue are seen small semi-transparent vesicles, which burst and leave grayish indolent ulcers. There is no vegetable growth on the surface, as in true thrush. Billard regards aphthæ as a disease of the mucous follicles characterized by a peculiar sebaceous deposit.

Occasionally the ulcers coalesce, but more often they die away, fresh ones appearing, and the disease, if unchecked, tends to run a very chronic course. In the majority of cases attention to the stomach and bowels will suffice to cure the disease. Again and again children have been brought to the Victoria Hospital for "ulcerated mouths," which, when examined, are found to contain the *fons et origo mali* inside them in the shape of a lump of unripe apple, some trashy sweet, or other stomach-deranging substance, with which the poor are forever "quieting" their children. Forbidding such and using the glycerine of borax application (better than Mel Boracis) soon cures the case; a little rhubarb and soda may be given to assist in the correction of the stomach and bowel disorder. Chlorate of potash is useful in the more severe cases, but its value will be specially considered under the next heading.

2. *Ulcerative Stomatitis, or Noma.*—This is a rapidly spreading ulceration of the gums, therein unlike follicular stomatitis, which spreads but slowly and destroys tissues but rarely. In this disease there is also heat of the mouth, submaxillary tenderness, restlessness, feverishness, and disordered bowels; the child is always putting its fingers to its mouth and picking at the throat. On examining the mouth irregular patches are seen of a dirty white grumous deposit; when this is removed the gum beneath is very red, raw, and bleeding at the least touch; the gums generally are swollen and spongy. If neglected, large sloughing ulcers form, the tongue becomes swollen and sodden, the saliva horribly offensive and diminished in quantity, and occasionally true gangrene of the mouth sets in, though this is rare. I have seen one case in which such, however, undoubtedly occurred. The causes of the disease are stated to be general cachexia, deficient nourishment, bad hygienic condition, especially damp; it is common in autumn. It is commonest in children between one and eight years of age. It also occurs after and in the course of lowering dis

eases, as the eruptive fevers, inflammations, &c.; occasionally it has prevailed as an epidemic. M. Taupin believes it to be contagious, *i.e.*, communicable by using the same spoon in feeding, &c.

Treatment.—These cases are usually easily cured by the use of chlorate of potash in good doses, given three times a day, the mouth being rinsed well with a weak solution of the chlorate, or in young children syringed therewith if the child be too young to rinse the mouth; glycerine of borax may be applied after each cleansing. The stomach and bowels must be regulated, and when improvement takes place tincture or decoction of bark will be of great use—in fact, the best form of tonic. Nutritious diet and wine are often necessary when the child is low.

Chloride of lime is used by M. Bouchut, forty-five grains of the chloride to six drachms of honey and application made with a camel's hair brush. Dilute nitric and hydrochloric acids are occasionally valuable in obstinate cases as local applications; acid and bark or ammonia and bark being given meanwhile internally. Dr. Dewees especially recommends—

℞ Cupri Sulph. .... gr. x.
  Pulv. Cinchon. Opt. .... ℨij.
  Pulv. Gum. Arab. .... ℨj.
  Mel. Commun. .... ℨij.
  Aq. Fontan. .... ℥iij.
Ft. applicatio.

The ulcerations to be touched twice a day with it.

3. *Gangrenous Stomatitis* (Cancrum Oris).—This is a most fatal disease; fortunately it is a rare one. It occurs oftenest in children from two to five years old (but it may occur up to twelve years), and debilitated by foregoing illness, especially fevers. The disease commences with fœtor of the breath, a discharge of offensive saliva, a hard red, shining swelling in the one cheek, not painful, but very tense. Inside the mouth at the corresponding point to this swelling will be found an excavated, jagged, unhealthy ulcer, covered with a brown slough; this ulcer is thoroughly phagedænic in character, its discharge is putrid to a degree, it involves the gums, loosens the teeth, destroys the tissues around, making frightful ravages and leaving cavities; if the child lives long enough, necrosis of the jaw may occur; the fœtor increases, but deglutition is generally not interfered with, and the child may eat up to the last.

Gangrene of the mouth can scarcely be mistaken for anything except perhaps malignant pustule. M. Baron draws the following distinction— malignant pustule commences on the exterior, affects the epidermis first, then the corpus mucosum, chorion, and subjacent parts; gangrene attacks the mucous membrane first, then the muscles, and lastly the skin.

The only method of *treatment* at all attended with hope of success is early and free destruction of the gangrenous parts with strong caustics— nitric acid, for example—an operation for which chloroform is needed; if the gangrene be not destroyed at one application the acid must be reapplied after a few hours. Heroic measures give the only chance of success; the acid must be used thoroughly applied by a piece of lint tied on a splint of wood, and the sound parts protected as far as practicable from its action. The actual cautery and the acid nitrate of mercury have been successfully used; probably nitric or muriatic acid are the most manageable and effective. The mouth should be washed or syringed out with warm water to which Condy's fluid has been added, or a weak solution of hydrochloric acid in decoction of chamomile or of bark, or chlorate of potash may be so employed dissolved in bark, or the Liquor Sodæ Chlorinatæ (℥j.— ℥xij.). The strength must be supported by liberal diet and wine, and by enemata of beef tea; carbonate of ammonia, chlorate of potash, and bark, are good internal remedies. Pneumonia is not an uncommon complication of cancrum oris; its advent should be watched for and carefully guarded against by avoidance of chills and draughts. The most thorough cleanliness, careful and unremitting attention, will be needed to ensure recovery even in hopeful cases of this terrible disease.

3. CATARRHAL PHARYNGITIS—SORE THROAT—ULCERATIONS IN THE THROAT.

Sore throat is not so common in children generally perhaps, as in adults, but some children are very liable to it. Exposure to cold and wet is the common cause. It is one of the many possible results of "taking cold." Occasionally, however, its origin is in connection with a disordered stomach.

The child complains of difficulty in swallowing, with some sense of heat and dryness about the throat. The voice is altered, becoming hoarse and husky. There is generally some swelling and tenderness about the angles of the jaw, and if the child be of strumous diathesis, the cervical glands will also be found enlarged in many cases. There is often more or less headache, pains in the limbs, and general feverishness. The mucous membrane of the throat appears red, dry, and rather shining. Then there is formed a varnish of secretion over the fauces, tonsils, and elsewhere; this is in patches and is hawked and spit out if the child be old enough—otherwise, swallowed.

Treatment.—Tincture of aconite, in drop doses at two-hour intervals internally, and a wet compress externally, will as a rule speedily cure sore throat. Chlorate of potash is valuable, as in all diseases of the throat, and a gargle may be made of it if the child be old enough; or it may be applied by the atomizer, or it may be given internally in moderately large doses.

In *chronic sore throat* glycerine of tannin is a most serviceable application painted on with a camel's hair brush two or three times a day. There are cases, however, where a solution of nitrate of silver, or tincture of iodine appear more beneficial. Tonics, such as syrup of the iodide of iron and Parrish's chemical food assist in the cure of these lingering sore throats.

Ulcerations of the throat may be superficial or follicular (which are deeper, irregular, and of variable size), or sloughing, &c. Chlorate of potash internally or locally, a touch with solid nitrate of silver, antiseptic gargles, or washes to be used with a syringe, or, better still, antiseptics in the form of spray, form the leading indications in the varieties of ulcerated throat. Sulpho-carbolate of soda is a nice preparation and is of special use in cases showing disposition to slough. The strength must be supported by concentrated nourishment by the mouth (for a quantity of fluid even cannot be taken), and in bad cases this should be supplemented by nourishing injections. A lump of ice is often grateful in the mouth and the water is slowly and almost unconsciously swallowed, preventing thirst. Port wine is to be given, by the mouth, if practicable, otherwise by the rectum, with other nutriments (see Dietary). In the rare cases in which dyspnœa occurs the practitioner must hold himself in readiness to perform tracheotomy, should the symptoms demand it.

4. Cynanche Parotidea—Parotitis—(Mumps)

is an acute contagious specific inflammation of one or both parotid glands; most frequently the left is first inflamed; occurs usually but once in a lifetime, and most frequently in children above five years of age. It is ushered in with a feverish cold, and then pain is felt, and swelling is perceived about the angle of the jaw; the swelling is exceedingly hard and painful, and extends often from beneath the ear along the neck to the chin. There is pain in mastication, in articulation, and in swallowing. The disease occasionally occurs epidemically; it usually reaches a height in three or four days, and then declines in severity and gradually disappears; or metastasis may occur, a remarkable feature in this disease, to be remembered and watched for, and such metastasis may be either to the brain, which is highly dangerous, and exhibits itself either in coma or delirium and may end fatally in a few hours, or to the mammæ in girls, and the testes in boys. These parts then become painful and swollen. Damp, whether in weather, clothing, or beds, appear to be a favorable factor in the causation of this disease; it is very rare for suppuration to occur.

Treatment.—The swelling should be fomented several times a day with a flannel wrung out of poppy and chamomile lotion, and a linseed-meal poultice applied occasionally. The bowels must be opened by laxatives; a dose of calomel and jalap is useful at the very commencement. Leeches are quite unnecessary, except metastasis to the brain occurs, when a few may be applied to the temples; the feet must be put into mustard pediluvia, and a brisk aperient given every three or four hours. It is not considered good practice, generally speaking, to solicit the return of the attack to the parotids. Should the testicles or mammæ be attacked, they will require fomentations and the same general treatment in the way of purgatives and derivatives.

Parotid bubo may occur in the course of typhus, measles, scarlatina, &c., and it may also follow erysipelas of the face. The general condition is usually adynamic and will demand support while the bubo is treated upon ordinary principles.

5. Tonsillitis, Quinsy, or Inflamed Sore Throat

is rare in children under twelve years. When it occurs the first symptoms are those of a cold with rigors, feverishness, flushed face, and a husky voice; by-and-by dysphagia is manifested, but dyspnœa seldom or never; the tongue becomes excessively coated; thirst is great. On examination one or both tonsils are seen to be enlarged, red, and inflamed; the uvula and pharynx generally swollen, and often œdematous; the difficulty in swallowing increases, there is running of saliva, and expectoration of thick mucus. Pain, darting along the course of the Eustachian tube to the ear, is complained of at each effort of swallowing; the inflammation in children more usually terminates in resolution or hypertrophy of the tonsils than in actual suppuration. If matter does form, it usually discharges inwardly, and is small in quantity. The more common result is a chronic enlargement of the tonsils, which renders the voice thick, and occasions loud snoring in sleep, with occasional violent and spasmodic fits of coughing; these are caused by swallowing the elongated uvula, the end of which tickles the mucous membrane of the windpipe and so brings on these convulsive fits of cough. The treatment of acute quinsy is sufficiently simple. It is often possible to avert an attack by a timely emetic and purgative, an astringent gargle, or, better, an inhalation of some medicated vapor, especially that of sulphurous acid. The value of aconite given in the usual way is often extremely marked in quinsy. "Caught at the commencement," says Dr. Ringer, "it rarely fails to succumb in twenty-four or forty-eight hours." At the same time a sinapism should be applied to the throat externally. When the inflammation is more advanced, inhalations of steam, impregnated with poppy vapor or other anodyne, will be useful in allaying pain. Linseed-meal poultices should be applied outwardly; the bowels should be kept gently open; washing the mouth with a solution of chlorate of potash, ℨ j. to ℨ viij., to which two or three drachms of Tinct. Kino may be advantageously added, will serve to cleanse away much of the viscid mucoid secretion. Internally two medicines have been much vaunted, the hydrochlorate of ammonia and guaiacum; the latter may be given in the following form:

℞ Acidi Citrici... gr. xv.
   Pot. Bicarb .... ℈ ij.
   Tinct. Guaiaci.. ℳ x— ℨ ss.
   Mucilaginis ad.. ℨ j.
Dum effervescend. quartâ quâque horâ.

The former in doses of three or four grains in water every four hours.

I prefer to give the chlorate of potash in ten grain doses, occasionally adding half a grain of iodide of potassium to each dose. Blisters are of little service, and should always be avoided in children's diseases when practicable. When the abscess has broken, warm water must be used to rinse out the mouth; warm poultices, frequently changed outside; the strength should be well maintained throughout with light nutritious diet, the disease being a very lowering one. Stimulants should not be allowed till the abscess breaks, when port wine is best. Sometimes it may be necessary to open the abscess; generally it is better practice not to do so. The situation, the strength of the patient, the general character of the case, will be the guides in this matter.

For chronic enlargement of the tonsils I know of no application so valuable as the daily (and of course careful) application of tincture of iodine to them. I have repeatedly used this means in children of all ages, from three upwards, and have rarely found it fail to give relief; at the same time the syrup of the iodide of iron should be administered. I have no faith in, and have never seen benefit from, repeated blisterings, application of caustic, or the use of gargles. If the disease will not yield to the iodine application, the removal of a portion of the glands must be thought of; this may be effected either by excision with the guillotine, or with a pair of forceps and a bistoury, or else by the use of potassa fusa in the manner recommended by Mr. W. J. Smith. When hæmorrhage follows the excision of the tonsil the plan of giving an emetic to control it by mechanical pressure according to the plan of Dr. Wharton Hood, is worth trying. A troublesome *elongated uvula*, exciting cough, &c., will require to be snipped, and the base touched with nitrate of silver if it show a disposition to bleed. A small piece is sufficient to remove as shrinking occurs more or less after the operation.

6. Retropharyngeal Abscess,

which is rather a rare disease, is diagnosed by the dyspnœa as well as dysphagia caused by a retracted state of the head, stiff neck, and difficult articulation. The dyspnœa, which is often alarming when the child is lying down, is eased by raising it to a sitting posture. In these cases it is almost always necessary to open the abscess, as spontaneous bursting is slow, and, indeed, rarely occurs. The abscess is readily felt on examination as a firm round tumor just beyond the base of the tongue, and generally in the middle line. This affection occurs not uncommonly in the course of scarlatina.

7. Dyspepsia.

In infancy this is manifested usually by vomiting; "he is sick, throws up all his food," is a constant trouble with the mothers of poor children, and most frequently such sickness is the result of some error in diet or feeding; for example, the child's stomach is over-filled, or it is weaned carelessly, or improperly fed, or the mother has taken improper food or over-exerted herself. Sometimes the milk returns as it was swallowed, sometimes it returns curdled and sour-smelling. Very often the sourness is caused by careless washing of the bottle in children brought up by hand. The bottle should be always cleansed most carefully with hot water, and thoroughly freed from sour and curdled milk. Again, vomiting may be symptomatic of disease elsewhere, as is often seen in the precursory stage of eruptive fevers, acute inflammations, &c., or it may be the result of spasm, as in hooping-cough, or it may be due to gastritis and intestinal disorder. Should prolonged dyspepsia occur, especially with frequent vomiting, the child grows emaciated, and such cases often end fatally, from gradual and increasing exhaustion.

There is a variety of dyspepsia called by Romberg *gastro-malacia*, which is so commonly observed in the out-patients' department of a child's hospital, *i.e.*, amongst the children of the poor and neglected, that it deserves a few words of separate consideration. Such a child has been brought up by hand, is often a "nurse-child," or perhaps what in grim travestie is denominated a "love-child;" it has been uniformly fed improperly and as uniformly physicked (within an inch of its life) with all the opiates, Steedman's powders, and carminatives that the ingenuity of its protectors has suggested. We are then told that the child is cross (and small wonder), peevish, restless at night, the tongue is covered with whitish yellow fur, a few patches of ulcerated stomatitis here and there in the mouth. There is constant diarrhœa, and the stools are green, and very offensive. The child is very thin, with a puckered wrinkled face, its breathing quick and short, its general aspect pitiable in the extreme. Such a case generally goes on from bad to worse; the child, too ill to be brought out, is confined to its cot, where it lies on its back and moans. The eyes will in this stage of affairs be found partly closed, and the abdomen tympanitic though apparently free from pain even on pressure; the exhaustion deepens down to death or a sudden onset of convulsions closes the scene. After death a gelatinous softening of the coats of the stomach and intestines is found, with no signs of inflammation, the softening appearing to depend on a diminished cohesion of the tissues, the result, in fact, of slow starvation.

It must never be forgotten that vomiting is one of the earliest symptoms of tubercular meningitis; the cause of this affection must therefore be always diligently sought. If the fault lies evidently with the mother, treatment must be directed to her; but if it be clearly the child that is in ill health, inquiries should be made as to dentition, as to the state and regularity of the evacuations, and as to the feeding. Attempts must also be made to restore tone to the stomach, and at the same time to diminish its work. Regular intervals for feeding must be ordered, the food if objectionable, changed; ass's milk and goat's milk are often useful in such cases.

Dr. Meigs records a case of a child aged three years where neither milk, bread, nor meat, could be taken, but the child digested raw oysters, soda biscuits, and rennet whey, and upon that diet lived two weeks, when it was able to take the white meat of chicken.

In extreme cases the stomach should be rested altogether by giving only a little iced water; if this is retained, adding to it some isinglass or milk; in the severest cases of all a milk bath or nutritious enemas may be used to nourish the infant while the stomach is absolutely rested for several hours, when the above method of cautious feeding may be attempted. The bowels will need correcting, and if acidity be present soda with rhubarb will be useful, or, if there be much sour diarrhœa, Mist. Cretæ with a grain or two of Pot. Bromid. two or three times a day; a small sinapism to the epigastrium is often useful, or a little chloroform on lint over the stomach, covered with a warm linseed poultice, rapidly produces external redness, and, indeed, requires watching lest it vesicate. In more chronic cases, in older children, similar hygienic conditions must be insisted on as to quantity and quality of food, exercise, &c. In such cases tonics are valuable, especially the mineral acids, and I prefer the nitromuriatic; a drop of this well sweetened, and with Tinct. Cinchon. co. or Inf. Calumbæ, answers well. Pepsine may be resorted to in obstinate cases, often with great benefit. It may be given in the form of pill or as wine. *Constipation* is a common concomitant of some forms of infantile dyspepsia; this must be met, not by irritant purgatives, but by chosen diet, a soap suppository, a little manna, or syrup of senna. A teaspoonful of olive oil occasionally, or Sodæ Pot. Tart. with Inf. Rhei and an aromatic answers admirably, and has the advantage of being a good stomachic tonic. It can hardly be too often reiterated that such cases are not generally suitable for mercurial treatment, though it would seem by the practice of many that the shrewd remark of Sir W. Jenner should be more widely known amongst practitioners :—" When you see a child, don't always think of gray powders." Not that gray powders and calomel are never to be given, even in dyspepsia; in cases of pale, clayey, offensive stools, nothing does so much good as a grain or two of calomel, followed by some gently alterative aperient; but simply that the plan of always dosing children with mercurials is most reprehensible.

A glass of cold water early in the morning, gentle frictions over the abdomen, and regular exercise, are other methods of relieving habitual constipation. Small doses of tincture of nux vomica are also of service in this respect, doubtless by giving tone to, and increasing the peristaltic action of, the bowel. I have found nux vomica most useful, especially in cases in which the lower bowel appeared to be at fault. It is very important, in girls especially, to combat a habit of constipation early in life. How much misery would often be saved through life, how much drastic medicine and pills avoided. A little care with diet in children, say from eight to ten or twelve years old, will soon bring matters right. For example, eating brown bread instead of white or "whole-meal" bread, or eating porridge for breakfast, with treacle instead of sugar if necessary. Fresh ripe fruit, stewed fruit, vegetables, and plenty of running about in the fresh air, are natural aperients, and worth all the drugs in Water Lane. A child should be taught to solicit an action of the bowels regularly at a certain time every day, and to report the fact of their *not* acting to the nurse or mother. In how few families can a ready answer be obtained as to when a particular child's bowels acted last. The child and the mother try to "think back," whereas, were the above simple plan adopted, any irregularity would be reported as a matter of course, and the mother would often be in possession of the very earliest indication of disorder, which rectified, might be the means of saving some threatened illness. Until people think it worth while to study apparent trifles like these, we must expect to see the revenues of quack pill-venders increase, and the health of our women in particular decline.

Infants are occasionally brought for vomiting of blood. Such cases are usually caused by the infant having sucked in the blood first from a cracked nipple; occasionally, however, true hæmatemesis occurs, either from general congestion of the abdominal organs, from some irregularity in the establishment of the circulation at birth; and occasionally children having passed or vomited blood recover and do well, and we are left in ignorance as to the cause. The administration of a grain or two of calomel to carry away the blackened motions and an occasional teaspoonful of iced water is, perhaps, the most rational treatment in such cases.

8. GASTRITIS

or inflammation of the stomach, may end in softening, ulceration, or gangrene; the inflammation itself is rare, and its symptoms obscure; vomiting is always present; pain, when present, is paroxysmal. There may be diarrhœa or constipation; there is usually tympanites, with thirst, restlessness, and fever; there is always some degree of tenderness at the epigastrium. Gastritis may, of course, be produced by the swallowing of irritant poisons, otherwise improper feeding seems its principal cause. It may occur as a secondary affection in the course of fever and inflammation, or as a sequela of stomatitis. There are no symptoms pathognomonic of softening.

Subacute gastritis probably exists more frequently than has been supposed; at any rate, it is not rare to find some of its traces after death, in congestion, thickening and hardening of the stomach-walls. The symptoms are those of aggravated dyspepsia—want of appetite, alternating with craving for food, pain after food; more or less epigastric tenderness and uneasiness; occasional vomiting; offensive and disordered motions; a pasty unhealthy look about the face, with often dark rings under the eyes.

Gastric catarrh, or mucous flux, is a common ailment with children; it is apt to be left as a sequela of some diseases, especially measles and hooping-cough; worms, dentition, and, indeed, any irritation of the alimentary canal readily excites this condition; giddiness and sickness, bilious vomiting, and disordered bowels, are prominent symptoms of it. The appetite is variable, generally bad, or craving for food may exist, and the food be rejected shortly after it is taken. The child wastes, looks pale and cachectic; the breath is offensive; the sleep broken, the bowels constipated one week and relaxed the next, the motions slimy, fetid, containing mucus, and sometimes a little blood. Bronchitis and pneumonia often insidiously complicate this condition.

The treatment of gastritis will divide itself into those cases caused by irritant poisons and those arising idiopathically or intercurrently. If a child has swallowed poison it should be made sick immediately; next oil and albumen are the best remedies usually to hand; then the special antidotes may be given if time permit. In idiopathic gastritis the feeding must be diligently looked after, and if faulty corrected; if dentition be present the gums will require attention; if stomatitis, glycerine of borax should be freely used, and chlorate of potash administered. A warm linseed poultice will ease the pain and tenderness over the epigastrium, or poppy stupes may be substituted. A small quantity of a cool drink with a lump of ice in it, if vomiting be very persistent; a dose of gr. j. of calomel with gr. ¼ of P. Ipecac. co. will be useful in a young child, and such may be repeated once or twice a day, according to the general condition; a child will often get its first sleep after the administration of the calomel and Dover's powder, and awake refreshed and better in every way. The diet should be thoroughly bland, yet supporting. Good beef-tea is useful, as a small quantity can be given, and it is highly nutritious. Isinglass may be added to the milk with advantage for infants. Great care will be required as convalescence is becoming established to avoid a relapse.

In chronic or subacute gastritis regulated diet, gentle aperients, avoidance of heating food and excitement of every description, sucking of ice, pepsine in small quantities twice a day before meals. Stomachic tonics with a small dose of alkali, as, for instance, bicarbonate of potash and infusion of gentian. Nux vomica is often of essential service in restoring tone when the more acute symptoms are passing off. In gastric catarrh a purge of calomel and jalap is desirable at the commencement of treatment, the more that by it not only is the intestinal canal cleared of the vitiated secretions, mucus, &c., with which it is often loaded, but also worms, if present, are generally passed, and a hint thereby afforded as to further treatment. The next indication is to restrain the mucous flux, which may be done by the use of bismuth, with a bitter infusion. Cascarilla or calumba answers very well. For diet, milk guarded with lime water or soda water; yolk of egg, good fresh meat, and avoidance of sweets, pastry and the farinacea. Salt-water baths, fresh air and exercise, are by no means to be neglected. An occasional purge will be required during treatment; the best in my opinion is compound jalap powder, with a quarter or half a grain of leptandrin, than which no drug more efficiently carries off the mucus, besides which its action upon the liver is useful, that organ often being sluggish in these cases. Change of air and ferruginous tonics will complete the cure.

9. ENTERALGIA (COLIC), ENTERITIS.

Colic is known by the child becoming suddenly fretful and drawing up its knees, crying, and then becoming quiet as if nothing had occurred. A discharge of flatus, which may be promoted by gentle friction with the hand over the back and stomach and by the administration of a little sweetened dill water, gives the greatest relief. Constipation, cold feet, and improper food are the usual causes of colic, which is occasionally a pretty severe affection, accompanied with piercing screams, flushed face, tense knotted abdomen, and considerable jactitation. A warm bath, fomentations to the stomach, and a mixture of aromatic spirit of ammonia, dill water, and some alkali—perhaps lime water is best—form suitable treatment. Castor oil is the best after aperient. Dr. Condie recommends, as specially valuable to overcome this costiveness of infants and to prevent recurrence of colic—

℞ Ext. Hyoscyam............................... gr. iv—vj.
Magnes. Calcin............................... gr. xxiv—xlviij.
Pulv. Ipecac............................... gr. ij—iij.
Ft. pulv. xij. Capt. j tertiis horis.

Enteritis, pur et simple, is not so common as gastro-enteritis or enteritis associated with inflammation of the large intestines also. Vomiting, diarrhœa, abdominal pain and tension increased by pressure, fever, tympanites, and a red dry tongue, are the prominent symptoms. The child lies on its back with its knees drawn up and with an anxious worn look, which is very characteristic of abdominal disease. The treatment is similar to that of gastritis—restricted diet, poppy stupes, cool drinks, a warm bath. Calomel and acetate of lead, combined with ipecacuanha and henbane, are recommended by Dr. Condie; the dose of calomel should be small.

℞ Calomelanos.
P. Ipecac., ana.............................. gr. ij.
Ext. Hyoscyam.............................. gr. iv.—vj.
Acetat. Plumbi.............................. gr. viij.—xij.
Ft. pil. xij. Capt. j tert. quâ. hôr.

It is, perhaps, scarcely necessary to add that the diet must be closely watched and guarded after the acute symptoms have passed away. Opiates are useful when much pain is present; they may be given by the mouth or as an enema, as recommended under dysentery.

10. Diarrhœa

may be simple or inflammatory, acute or chronic, feculent, bilious, mucous, chylous, or enteric. It is better to consider the disease generally, and to notice the prevalence of special forms in the course of such consideration. In medicine names and varieties are comparatively worthless, as we so seldom meet practically with the typical forms. Diarrhœa is an important affection, because it is the very commonest ailment of infants and children. To recognize its presence we should remember the normal condition of the bowels in infants. A child should have from three to six motions in the twenty-four hours, the color of these should be deep yellow, the consistence that of thick gruel, and odorless, and much resembling a mess of mustard. From this standard, however, the departure may be considerable, and yet not inconsistent with health, and should not be interfered with. During dentition especially a little diarrhœa is frequently beneficial rather than otherwise. When, however, moderate limits are evidently passed, and especially if there be pain, manifested by the drawing up of the legs towards the abdomen, it will be time to interfere, and the first thing to do is to remove the cause of the ailment; this will often be found in the feeding. If the milk disagree, the nurse must be changed, or the mother's health corrected if irregular; or if weaned, the quantity and quality of the food must be carefully regulated. Broths are to be absolutely forbidden. I have seen many cases of diarrhœa kept up by the mother persisting in giving a child beef-tea contrary to advice. Rice and arrowroot are the main aliments to be employed in these irritations of the alimentary canal. Occasionally raw meat chopped small and finely shredded will be extremely beneficial. When resorted to it should be the only food given; it renders the fæcal discharges extremely fetid, but the benefit is undoubted; in many cases it has saved life. Milk should be diluted with lime-water and given cold; yolk of egg is unobjectionable. Sometimes a few drops of brandy may be added with advantage to equal portions of milk and lime water.

If the gums be tumid and tender they must be lanced, but not otherwise; it is folly to lance a child's gums on every provocation. If the diarrhœa be feculent a small dose of castor oil often carries off the mischief; if, on the other hand, it is frequent, watery and griping in character, chalk mixture, or aromatic confection, or small doses of Dover's powder, are the remedies. Decoction of logwood is a good medicine, but stains the linen; this should be remembered, as mothers often protest. Catechu and Tinct. Kino are valuable astringents in obstinate cases. In bilious diarrhœa a little gray powder will be needed instead of castor oil, and then the chalk mixture may be given, and it will be useful to add a little bicarbonate of soda in such cases. In mucous diarrhœa, when the stools are like chopped spinach, and occasionally have a little blood in them, P. Ipecac. co. is useful, and tends to soothe the tenesmus commonly present. Ipecacuanha is a drug the value of which in many forms of diarrhœa is indisputable. Small and repeated doses act best. When the discharges are white and milky looking, or containing apparent particles of fat (so-called chylous diarrhœa), small doses of calomel and opium will be required, or gray powder in very small doses by itself will often correct matters. Raw meat finely shredded is especially useful in these cases, as the child gets rapidly emaciated. Enemas containing a few drops of some anodyne are also of value. In lienteric diarrhœa, in which the food passes through unchanged, it is evident that the stomach is mainly at fault, and our efforts must be directed to correct its condition; the amount and quality of the food must be carefully regulated, the mineral acids, or Syr. Ferri Phosph. co., or reduced iron with pepsine, may be given. The child should have salt-water baths and be well rubbed after with a Turkish towel, and should sleep in a large well-ventilated room; it should have an abundance of fresh air, and moderate exercise.

Dr. Müller regards nux vomica as a specific in this form, and remarks —as, indeed, the nature of the disease fully demonstrates—that opium is useless.

In acute catarrhal diarrhœa, summer diarrhœa, and the diarrhœa of dentition, nitrate of silver in doses of from $\frac{1}{8}$—$\frac{1}{14}$ grain is useful. Especial indications for its employment are according to Dr. Müller—(1) croupous deposits on the mouth and fauces; (2) peculiar redness and smoothness of the tongue; (3) irrepressible thirst.

In chronic diarrhœa, especially of the feculent variety, I have found the greatest benefit from the use of carbolic acid (Calvert's) internally in doses of gr. $\frac{1}{4}$ to gr. $\frac{3}{4}$ in well-sweetened water for children about two years old. I have sometimes combined with this a little of the Syr. Ferri Phosph. co. with great advantage. The intertrigo which occurs from acrid diarrhœa will be best treated with copious warm-water ablutions, and subsequent dusting with pure oxide of zinc or lycopodium, which in bad cases is preferable, as water glides over its surface as if over oiled silk. The following aphorisms of Bouchut are eminently practical comments on the varieties of diarrhœa in childhood.

1. Yellowish homogeneous diarrhœa is generally of little importance.

2. Yellowish diarrhœa becoming green on exposure to the air under the influence of the action of the urine is unimportant (this is the so-called suburral diarrhœa).

3. Yellowish-green diarrhœa, or that sprinkled with specks of curd, indicate considerable intestinal irritation.

4. Abundant serous diarrhœa is always an unfavorable phenomenon.

5. Catarrhal diarrhœa sometimes engenders inflammation of the intestines.

6. Diarrhœa leads to enlarged belly amongst children.

Besides the ordinary forms of acute or subacute diarrhœa there is also chronic diarrhœa, occurring after the acute specific diseases, or bronchitis, or pneumonia. Rilliet and Barthez records 140 cases of secondary chronic diarrhœa; 37 were preceded by measles, 27 by pneumonia, 17 by typhoid, variola, and scarlatina, and 29 by other diseases, including bronchitis, croup, pleurisy, &c. Of this number 21 only recovered, all the rest died. There is an insidiousness about this form of diarrhœa; it is a little better one day or week, and a little worse another, so that it is not unfrequently overlooked or neglected by the parents. Chronic diarrhœa may be kept up or caused by worms, by improper diet, by neglect and bad hygienic conditions generally, especially by cold or sudden changes of temperature. Lastly, in children over two years of age chronic diarrhœa may be tubercular in origin; this is notably the case in tubercular peritonitis and in tabes. The general history and presence of tubercle elsewhere will guide the diagnosis. In all these cases there is considerable resemblance in the general symptoms produced—the child is pale, leaden colored, his skin harsh, dry and burning, lips and eyelids anæmic, with a dark line under the eyes; there is gradual but steady emaciation; capricious appetite, at one time ravenous in character, at another rejecting everything offered. The abdomen is tumid and generally tender on manipulation; the stools gradually increase in frequency from three or four to ten or fifteen in the twenty-four hours, with tenesmus and occasionally streaks of blood. The stools become more and more offensive and unnatural in character, mucus appears in quantity; sometimes they are green for a time and then turn clayey and pasty. The tongue is very characteristic, red and glazed, rarely furred. As the disease goes on there may be a little œdema or actual serous effusion into the pleura, and more rarely into the peritoneum. Besides which, if the child keep its bed, pneumonia—the pneumonia of stagnation—is almost sure to occur, without cough, and recognizable only by the physical signs. When the mouth becomes aphthous and the cry ceases to have a good "return," but passes on to a wail, the end is close at hand.

Treatment will comprise, first and foremost, due attention to all hygienic matters, especially food and warmth; the feet are by all means to be kept warm, and flannel should be worn next the skin. A flannel roller should be wound round the abdomen of young children suffering from diarrhœa. The parents are to be impressed with the vital importance of regular and judicious feeding. Rice, arrowroot, and baked flour will be useful in young children; the quantity given at a time should be moderate, that the stomach and intestines be not overloaded. Milk guarded with a little brandy is useful. The stools at first being generally offensive as well as frequent, a powder containing a small dose of gray powder, with a little rhubarb and carbonate of soda, is appropriate medicine. In some cases a grain of calomel with sugar may be better, but in no case is the mercury to be continued, *however great benefit* result from its single use. In a rickety or scrofulous child it is better not to give it at all, nor yet where tubercular depositions are suspected. Rhubarb, soda, and a little ginger, will suffice in such. Afterwards it is a matter of some nicety to select the best remedies; the greatest care is necessary lest mischief rather than good be done. If there be any leading symptom it is sound practice to take it as a guide, at the same time not to be guided by it slavishly, *e.g.*, much griping and tenesmus will be benefited by an opiate.

℞ Tinct. Opii.................................... ♏j.—ij.
   Pot. Citrat..................................... gr. iij.—vj.
   Syrupi Aurantii................................ ℨss.
   Aq. Cinnamom.................................. ℨij.
Ft. M. tert. vel quárt. q. hor.

If this fail, an enema of starch and a few drops (three or four) of laudanum may be employed. Bismuth one or two grains frequently, or chalk, are indicated by sour-smelling frequent stools, castor oil when there is offensiveness and the motions are green. Dr. West's prescription given under the head of dysentery is the nicest form in which to administer it. Astringents are only to be used to check *frequency;* red gum, krameria, and catechu are, perhaps, the best. In obstinate cases Trousseau recommends an enema of gr. j. nitrate of silver to ℨv. of water twice in the twenty-four hours; a mixture of chalk, bismuth, and mucilage, with a few drops of Tinct. Opii, or, better, Liquor Opii Sedativ., being also given every few hours. Oxide of zinc has been recommended by Dr. Brakenridge, two grains for a child a year old in a little sweetened mucilage and water every two or three hours; this remedy is specially directed to the debilitated state of the nervous system in children with relaxed and irritable bowels. I have tried oxide of zinc in a few cases, twice to be much pleased with the result, but at other trials to be disappointed. In one of these cases where oxide of zinc signally failed, a few small doses of veratrum album were quite successful. I have known this remedy to do good in other cases, and intend to give it a further trial. Camphor is often useful in the sharp diarrhœa of infants; a few drops on sugar, or in milk, of the spirits of camphor, may be given every two or three hours. Lastly, the use of arsenic is advocated by Dr. Ringer in cases of children from eight to twelve years, where the motions are semi-solid, but contain lumps of half-digested food. One or two drops of the Liquor Arsenicalis before each meal is the dose, and the same author recommends a minute dose of the perchloride of mercury "for very *slimy* stools, especially if mixed with blood, and accompanied by pain and straining." It is this slime, colored with a small quantity of blood and all coagulated together, that forms the "lumps of flesh" which mothers constantly describe their children as passing. Dr. Ringer recommends one grain of the perchloride in ten ounces of water, a teaspoonful every hour. For my own part I have found benefit in such cases, sometimes from tincture of nux vomica, and sometimes from a combination more or less nearly approaching the following of Dr. T. King Chambers:

℞ Tinct. Opii.................................... ♏x.
   Pulv. Ipecac.................................. gr. ij.
   Sodæ Bicarb.................................. gr. xx.
   Theriaci...................................... ℨij.
   Aquæ......................................ad ℨiss.
Ft. Mist. Dose, ℨj. (one teaspoonful) every hour.

11. Dysentery, or Inflammatory Diarrhœa

is a condition so important and so much more fatal than the ordinary form of diarrhœa as to require separate and careful consideration. Dysentery may arise idiopathically or as a sequela of protracted diarrhœa. It is usually preceded by vomiting accompanied with frequent purging, sometimes almost incessant, till the stools, at first natural, become slimy, and then streaked with blood; their expulsion is accompanied with straining and tenesmus more or less severe. After a while no relief is obtained by the passage of the stools; on the contrary, the pain and tenesmus are increased; the abdomen now swells, is tender and burning; there is general fever and restlessness, rapid emaciation, the discharge increases in offensiveness, the parts round the anus becoming excoriated; aphthæ and cancrum oris, bronchitis and cerebral irritation, are common complications towards the close of the disorder.

The causes of dysentery, when the disease is not intercurrent in the course of fever, appear to be especially damp, heat, and foul air; to these bad feeding, insufficient clothing, and dentition may be added. It is a common epidemic in hot climates, and occasionally it prevails as such in northern latitudes. The disease being essentially inflammation and ulceration of the colon, it is usual to find the evidences of these processes after death. The mesenteric glands are often enlarged, but otherwise unaltered in color or texture. The small intestines are generally healthy, the liver is often gorged, and it is common to find the mucous membrane near the ilio-cæcal valve congested and thickened. It is also common to find the mucous membrane on the rugæ destroyed, especially in the sigmoid flexure of the colon and in the rectum. It is always necessary to examine the rectum carefully for evidences of ulceration, as such are frequently confined to that portion of the bowel. The prognosis of dysentery, even when idiopathic, should be guarded. It is eminently a dangerous disease.

Treatment.—The child should have a warm bath, and subsequently a bran or linseed poultice should be applied over the abdomen. It is a good plan to make such a poultice with strong decoction of poppies instead of water. If the vomiting will permit of the administration of medicine, the treatment recommended by Dr. West is often beneficial, viz.—

℞ Ol. Ricini.. ℨj.
   Pulv. Acaciæ.. ℈j.
   Syrupi... ℨj.
   Tinct. Opii.. ♏iv.
   Aquæ Flor. Aurant.. ℨvj.
   Misce.—A teaspoonful every four hours for a child one year old.

If such, however, be rejected, an enema consisting of half an ounce, or less, of mucilage or starch, with three or four drops of Tinct. Opii should be carefully injected, which frequently relieves the tenesmus very speedily. Mucilage and chalk mixture, with opium or P. Ipecac. co., in small doses, are also useful. Ipecacuanha stands out as the drug perhaps most generally serviceable in dysentery; by some it is regarded almost as a specific. Fair doses should be given, and the powder appears the best form. Throughout, the strength must be well supported, and brandy is of all stimulants the best; its need must be judged of from the general circumstances of the case. Raw meat finely shredded will prove of the utmost use in supporting the strength, and small quantities of strong extract of meat may be given from time to time if the strength flag.

Milk, arrowroot, and rice may be freely given. When the more acute symptoms pass away is the time for astringents, the aromatic sulphuric acid, with tincture of bark, Tinct, Kino, hæmatoxylum, catechu, red gum or lead, will be of service. The child will require care and watching for some time after all apparent danger is over, from the tendency of slight causes to bring on a recurrence of the symptoms. Minute doses of nitrate of silver are recommended by Trousseau when the disease shows a tendency to become chronic. Sulphate of copper, tannin, and iron baths may also be used under such circumstances, or the following mixture given:

R Liq. Ferri Pernitrat.. ʒ ss.
Acidi Nit. dil.. ʒ ss.
Syrupi Zingib... ʒj.
Aq. Anethi,...ad ʒiij.
ʒ ij. every six hours.

Small doses of the perchloride of mercury are recommended by Dr. Ringer, and the tincture of hamamelis is useful, especially if there be much blood in the stools.

Since writing the above I have had an opportunity of treating a well-marked case of very chronic dysentery contracted in the Fiji islands in a child about fourteen years of age. Her condition was one of much anæmia—a considerable daily loss of blood—very frequent stools, often of the "lumps-of-flesh" character. This young girl had been under homœopathic treatment for some time "and was nothing bettered, but rather grew worse." Restricting her to suitable diet, I gave five drops of the solution of perchloride of mercury with a drachm of tincture of bark three times daily—giving four grains of Dover's powder. The improvement within a week was so remarkable that the mother seemed to think that perhaps too much had been effected. But the young lady being able to take several trips by steamer, of a rather trying character during extremely hot weather, without ill-result, was a sufficient indication of the real benefit attained.

12. Worms.

We infer that a child suffers from worms when we hear that it picks its nose, grinds its teeth at night, has a voracious and capricious appetite, looks pinched and thin at the bridge of the nose and dark round the mouth, has a tumid belly, dark rings round the eyes, is often sick, and complaining of itching at the anus. Dr. Eustace Smith describes the tongue of worms as slimy, thickly furred, with enlarged papillæ at the side. But none of these symptoms are actually diagnostic; *seeing* the worms or parts of them pass away is the only infallible indication of their presence. It should be remembered that scrofulous children are the most likely to be infested with worms. The varieties of worms are:

I. Cœlelmintha, or hollow worms (having an abdominal cavity).

1. *Trichocephalus dispar* (long thread-worm), two inches long, slender. Habitat, cæcum and large intestines.

2. *Ascaris lumbricoides* (round worm), like a common garden worm; light yellow, three to nine inches. Habitat, small intestines; is occasionally vomited up from the stomach, into which it has crawled.

3. *Ascarus* or *Oxyuris vermicularis* (thread-worm), quarter of an inch

long; very common. Habitat, cæcum, whole length of colon, and rectum; often comes away in masses.

II. Sterelmintha (solid worms), having no abdominal cavity.

1. *Tænia solium* (tape-worm), five to ten feet in length; common. Habitat, small intestines; white and flat; the head small, armed with four suckers, between which is the mouth surrounded with five hooks. Rare in children under six years.

2. *Bothriocephalus latus* (broad tape-worm), twenty to one hundred feet. Habitat, small intestines; its segments are broader and shorter than those of the common tape-worm; it is almost confined to the inhabitants of Switzerland, Poland, and Russia.

Treatment.—1. Of the worms which chiefly infest the lower bowel.

Ascarides.—Injections of salt and water, of quassia, of santonine, gr. iv.—gr. viij. to ʒ iv. lime water, or lime water ʒ iv. to which 3 j. or 3 iss. of Tinct. Ferri Perchlor. has been added. With such an enema should be combined an aperient powder of calomel and scammony or jalap. Dr. Spencer Cobbold asserts that the cæcum is the true headquarters of the thread-worm, and that therefore remedies by the mouth are as necessary as those by the bowel. With this view he employs saline cathartics with or without aloes and sulphate of iron. Bitters of all kinds are useful. The infusions should be drunk in good quantities rather than concentrated forms employed, even after numbers of worms have been got rid of; the child should take some Tinct. Ferri Perchlor. in quassia infusion for a while, and be directed to have its food well cooked and eaten with plenty of salt.

Trichocephalus dispar, or long thread-worms, must be treated on precisely similar principles.

2. Treatment of those worms which infest the small intestines.

Round worms.—Santonine in dose of from two to four grains may be given at night (the child having no supper, the better to expose the worms to the action of the drug), followed by a teaspoonful or two of castor oil or other aperient the next morning before breakfast. The process may be repeated for a night or two until the canal seems cleared. Santonine occasionally produces giddiness, sickness, and disordered vision; it is often given in the form of a lozenge or worm-cake.

New Zealand being a country where round and thread-worms are *extremely* prevalent in children, and, indeed, in adults also, at any rate in the north island, I have had pretty good opportunities of testing santonine and other remedies, and can safely say that so far I have found none more effectual than santonine, nor can I remember in really hundreds of cases of its employment one in which disagreeable symptoms were manifested. I have tried all ways of giving it, by itself, with jalap or scammony, and I think there is none better than with an equal dose of calomel, according to the age of the child, from one to three grains of each, Quassia and steel, as above indicated, I employ during treatment and for some time after all worms are considered expelled. The concentrated tincture of chelonin, about five drops, repeated if necessary, is also a valuable remedy against round and thread-worms.

Spigelia or Carolina Pink may be thus given:

℞ Pulv. Spigeliæ... gr. x.
   Pulv. Stanni.. ʒ ij.
   Syrup. Zingiberis.. ʒ j.
   Mellis... q. s.

(*Neligan.*)

This makes a paste or bolus, which may be taken at bedtime and followed in the morning by a purge.

℞ Ol. Chenopodii.. ʒj.
    Pulv. Acaciæ... ʒij.
    Syrupi.. ℥j.
    Aquæ Cinnamomi... ℥ij.
Ft. M. Coch. j. med. t. d. for three days, and again repeat after several days.

Mucuna Pruriens, or cowhage, acts only as a mechanical irritant; it is seldom used. The dose is a drachm with honey before breakfast, followed by castor oil.

Tape-worms.—Against these we have many remedies. Oil of male fern, thirty to sixty drops of the liquid extract (Ph. B.) in mucilage, fasting. Of the old ethereal oil, ten to thirty drops. A purgative should be administered before and after the male fern. This drug is occasionally successful against round worms.

Turpentine combined with castor oil in rather large doses is a good remedy; the dose varies from ʒss to ʒij. This drug also expels round worms.

Decoction of pomegranate root bark, in ounce doses, is an effectual but bulky and unpleasant remedy.

Kousso in ʒj.—ʒij. doses, is also efficacious; it requires an aperient subsequently. Petroleum and creosote have also been recommended as successful against tape-worms; of petroleum ℳx. or more, of creosote ℳj., two or three times a day.

Kamala, twenty to sixty grains of the powder, or in the form of tincture:

℞ Tinct. Kamalæ................................ .... ʒss—ʒj.
    Syr. Aurantii.. ʒj.
    Mucil. Acaciæ... ℥ss.
Ft. haustus.

To be followed, as in the case of most anthelmintics, by a purge.

After any worm expulsion ferruginous and bitter tonics should be administered for a few weeks.

It must not be lost sight of that symptoms sufficiently alarming are often due to the presence and irritation of worms—for instance, diarrhœa, dysenteric in character, with much tenesmus; frequent, painful and involuntary micturition; obstinate leucorrhœa, convulsions, vertigo, actual syncope, perversions of sight, and sometimes temporary blindness, strabismus, &c. Of symptoms considered especially characteristic of the different kinds of worms may be enumerated—itching of the anus of oxyurides; tenesmus and vertigo of round worms; abdominal pain, gnawing in character and accompanied with some swelling, especially about the umbilicus, of tape-worm. When, as occasionally happens, considerable dyspeptic disorder or intestinal irritation has been set up by the long-continued presence of worms, the diet will require careful regulation, moderation in vegetables, especially potatoes, and abstinence from farinaceous food; appropriate stomachic tonics, combined with an occasional aperient, should be persevered in until all symptoms are removed. Worms are probably only actually dangerous to life when they emigrate to neighboring and important organs, *e.g.*, the air-passages or liver.

13. Jaundice.

Icterus neonatorum is jaundice occurring within a few days of birth, and disappearing, usually spontaneously, in a week or so. It occurs most commonly in infants prematurely born and feeble in constitution, and seems to be dependent on defective respiration and impaired performance of the functions of the skin, to which the hepatic disorder is but secondary (West). The treatment of these cases is most simple; careful avoidance of chill, a mild aperient, mercurial if necessary, and keeping the infant rigorously to the breast, suffice for their cure. Occasionally the jaundice of infants is more serious in origin; it may depend on congenital absence of the hepatic or cystic bile-ducts; in such cases bleeding readily occurs from the umbilicus, and, being destitute of coagulative properties, it is controlled neither by styptics nor ligature, but a constant oozing goes on from the granulating surface of the navel. When the bleeding, however, is controlled (e.g., by transfixing the integument at the root of the navel by a couple of hare-lip pins, and twisting round them a strong silk ligature), death stills usually occurs from atrophy or exhausting diarrhœa in a week or two. Jaundice in older children is caused, as in adults, by some temporary impediment to the flow of bile into the duodenum, and by defective secretion on the part of the blood. In both cases biliary matters, not being separated from the blood, get into the general circulation. The ducts may be obstructed by gall-stones, which are very rare in children; by cancer of the liver or pancreas, by inflammation, spasm, and even by constipation, the loaded bowel pressing on the ducts. On the other hand, the elimination of bile may be checked by congestion and inflammation of the liver, by mental emotions, and by gastric disorders.

When jaundice is established the skin is saffron yellow, so are the conjunctivæ; the motions are pale, the urine dark, and there is usually pain or weight in the liver region. The skin is often dry, there is bilious vomiting, headache, and vertigo, restlessness, sleeplessness, and morose or uncertain temper. Such cases yield readily to small doses of gray powder at bedtime, a little liquid extract of taraxacum with a small quantity of Epsom salts by day, and when recovery is progressing small doses of nitro-muriatic acid will complete the cure. Leptandrin and nitro-muriatic acid are remedies most valuable in the more chronic disorders of the liver in young children. Such have often been well dosed with calomel and gray powder, and it is then, by all means, desirable to avoid irritating the liver and damaging the general health further by their exhibition. If purgatives be required, it will be better in such cases to give a small dose of leptandrin, gr. $\frac{1}{16}-\frac{1}{8}$ with podophyllin gr. $\frac{1}{8}$, which for a child three or four years old will act admirably; these powders may be repeated twice or three times a week, when necessary, to promote the proper secretion of bile, and with their use it will scarcely be necessary to employ mercury in any shape. Nitro-muriatic acid baths are also serviceable in chronic cases. Dr. Waring Curran thinks the bile is retained in many cases by a sort of spasmodic retention, for which he recommends tincture of belladonna in two-drop doses.

14. Acute Peritonitis,

rare idiopathically in the adult, is rarer still in the child; our notice of it, therefore, need be but brief. It is known to occur occasionally before birth, and to cause the death of the fœtus; such cases are probably syphilitic in origin.

The most prominent symptoms of peritonitis are pain aggravated by pressure and movement; even the weight of the clothes is unbearable; the child lies on its back, with its knees drawn up, with acute suffering stamped on its features; the abdomen is swollen, often tympanitic; there is restlessness and fever, diarrhœa oftener than constipation, and sometimes nausea and vomiting; a thready, rapid, jerking pulse. When effusion of serum takes place the belly ceases to be tympanitic, but may again become enormously so shortly before death. The symptoms of secondary peritonitis are precisely similar, though perhaps less intense; it occurs secondarily in the course of ascites, typhoid, from perforation of the bowel, scarlatina, erysipelas, &c. The prognosis is always grave. Dr. Churchill says, "there is no more mortal disease."

Treatment.—This should consist in relieving the pain by anodyne fomentations, or covering the stomach with extract of belladonna made soft with a little glycerine, then calomel and opium should be administered guardedly, and anodyne enema may be resorted to, to procure sleep. If there be much diarrhœa the quantity of calomel must be lessened, and the quantity of opium increased. It is well in a child two years old to begin with half a grain of calomel, and one grain of P. Cretæ cum Opio every two hours; in no case can good result from cathartic purging. Blisters are also inadmissible, and in my opinion bleeding in any form. If constipation happens to be present a lavement will relieve it.

The diet should be low at first, but the child must be seen every four hours, and the indications for improved nourishment and even stimulants narrowly watched for.

Ice is useful in controlling vomiting when present. Some practitioners recommend the use of mercurial ointment rubbed into the thighs to hasten salivation; ℥ ij. of Ung. Hydrarg. may be spread on lint or muslin and laid on the abdomen, or rubbed into the thighs. The attack may degenerate into *chronic peritonitis*, which disease, however, after three years of age, when it occurs most frequently, more often arises independently; it is almost always associated with tubercle (and is hence called *tubercular peritonitis*). The attack is insidious; usually there is diarrhœa, painful or painless; the child seems fairly well otherwise. Later come pain and sensation of tightness in the abdomen, which is found swollen; the pain then becomes more frequent and paroxysmal in character; obscure fluctuation is then perceptible, and the veins over the belly are large and prominent.

Pain in tubercular peritonitis is rarely constant or confined to the seat of lesion; it is rather shifting, intermittent, colicky pain, and there is usually tenderness on pressure over the abdomen; and the fluctuation is not due usually, if ever, to ascites, but to the transmission of the manual impulse by the agglutinated intestinal masses (Rilliet and Barthez). When tumor can be detected the omentum is almost always the site of it.

Still, the child's tongue may be tolerably clean and its appetite fair, and the bowels not very irregular. As effusion goes on, dyspnœa sets in, the pulse quickens, the skin gets hot, there are exacerbations morning and evening, loss of strength, great emaciation, diarrhœa, and death from exhaustion or some intercurrent tubercular affection rapidly follows.

The treatment will comprise, most especially, the use of iodide of potassium, and iodine ointment with cod-liver oil is to be rubbed frequently into the abdomen, which may also be painted with iodine-paint, and over which warm applications may be laid to ease pain when required. Wolff recommends powdered digitalis with bitartrate of potash in small doses. The diet must be nourishing and the strength supported in every way. Sea air is occasionally beneficial, and tonics are useful so soon as the disease shows any sign of yielding, but it is unfortunately sadly fatal under any treatment.

15. Tabes Mesenterica

or tubercular degeneration of the mesenteric glands, resembles chronic or tubercular peritonitis very closely in its symptoms. It is rather a rare affection. Children from the twelfth month to the eighth year are most liable to it, but it is especially rare before the fifth year.

The presence of small quantities of tubercle in the mesenteric glands is, however, exceedingly common.

The symptoms are pain, more or less persistent, in the abdominal region, so that the child lies at such times with its legs drawn up to the belly; the bowels are irregular, confined or relaxed, often clayey and offensive; the abdomen remains large and tumid, whilst the rest of the body wastes (and this will distinguish such enlargement from that of rickets, e.g.), and debility rapidly increases. The disease cannot be diagnosed with absolute certainty until the enlarged glands can be felt through the abdominal walls, and this never happens until the disease has progressed for some time.

As the glands become enlarged various symptoms referable to pressure upon neighboring organs are manifested. Œdema and enlargement of the abdominal veins are common. Ascites is commonly, but not always, present; the mechanism of its production may be either pressure upon the thoracic duct or vena portæ, or chronic peritonitis occurring at various times in the progress of the disease; such peritonitis is rarely general, it is more often local, and post-mortem patches of recent and remote origin may be found. In such cases pain, fever, and the other ordinary symptoms of peritonitis, may be absent or so obscure as to attract no attention.* Occasionally the ascites is very considerable, and fluctuation is very distinct; it is, perhaps, more often small in quantity, and the fluctuation indistinct from adhesions. There may be considerable tympanites, occasionally vomiting and diarrhœa, the tongue usually pretty clean, and the appetite good. Cases are seen in which no swelling whatever of the abdomen occurs; it is, on the contrary, rather shrunken; in such cases the tumor can usually be felt easily; it is hard, roughly nodulated, and somewhat movable. The tubercular deposits of tubercular peritonitis, e.g., in the omentum, &c., are generally softer, less nodulated and more movable. Fever, pain, early abdominal swelling and tenderness, with more or less obscure fluctuation, as prominent symptoms, point to peritonitis; but inasmuch as the two diseases may coexist, the diagnosis is often very difficult. Considerable hypertrophy of the cervical glands will indicate the probability of tabes rather than peritonitis.

Hectic fever sets in towards the close, the pulse becomes extremely rapid, and profuse sweats are common; the child dies at last either of exhaustion, of some intercurrent attack of enteritis, or peritonitis. The disease frequently lasts for months, with exacerbations and remissions, and it does occasionally, but very rarely, terminate in recovery.

---

* See a case reported in the "Lancet" of August 21st, 1869.

The *treatment* will be mainly that of tubercle elsewhere, viz., supporting the strength, correcting the disordered state of the liver and bowels, and the administration of iodide or phosphate of iron and cod-liver oil. Inunction of iodine ointment and cod-liver oil is useful. Change of air, especially to sea air, is all important, and bathing in warm or cold sea water, according to the season of year, should be resorted to. Raw meat, cream, chocolate, and cocoa, are valuable nutriments in these cases. Dr. Dobell recommends pancreatic emulsion in teaspoonful doses every four hours.

16. Abdominal Tumors.

It must be remembered that when children are brought "for a big belly," it very often happens that the abdominal enlargement is only what is really quite natural and wholly unconnected with disease of any kind. Children naturally have "big bellies." In other cases the enlargement is evidently associated with rickets or tubercular peritonitis.

The liver is the organ which most frequently gives rise to genuine abdominal tumors in children; after the liver the spleen, and after the spleen the kidneys, are the glands most commonly at fault.

The liver may be enlarged from fat deposit, constituting "*fatty liver.*" Frerichs calls attention to the frequent association between fatty liver and pulmonary tubercle. In infants at the breast the hepatic cells are usually rich in fat, and the liver is often to be felt enlarged during the first few weeks of life from injudicious stuffing. Again, *hyperæmia* and enlargement of the liver may result from obstruction to the circulation of the blood, as in morbus cordis, especially in mitral disease, both obstructive and regurgitant, and also in constriction of the thoracic cavity from angular curvature. In such cases of mechanical obstruction to the onward flow of the blood the liver becomes uniformly enlarged, its capsule distended, and its parenchyma firmer. On section it presents a "*nutmeg*" appearance, and hence this condition is often called nutmeg liver. The symptoms are (besides those to be ascertained on manipulation and percussion) those of gastric catarrh, and also slight jaundice, constipation, and scanty high-colored urine, slightly albuminous. The enlarged organ at first feels smooth and tense, later it is uneven and granular. The treatment must be mainly palliative, inasmuch as the *fons et origo mali* is incurable. When much pain exists in the hypochondrium a few leeches may be useful, or dry cupping. Mild saline purgatives, with taraxacum and the occasional use of a leptandrin or podophyllin powder, are the chief indications. Frerichs recommends the waters of Marienbad and Kissengen.

Amyloid, albuminoid, or *waxy* liver, gives rise to considerable tumors. It is associated most frequently either with (a) caries or necrosis of the bones, especially in the joints and vertebræ, more rarely in the shafts of the long bones; (β) rickets; (γ) constitutional syphilis; (δ) ague; (ϵ) tubercle of lungs or intestines.

The disease usually lasts over many months; its time of origin is, however, rarely noted, as it gives rise to few symptoms during its early process, and, indeed, the liver may attain considerable size before the general health is at all interfered with; by-and-by, however, albuminuria, ascites, and enlargement of the spleen are noticed. The liver itself is hard, the surface smooth, the edge sharp and well defined; these points, the tumefaction of the spleen, and the association with caries, struma and syphilis, are diagnostic from hyperæmic swelling, and, indeed, from fatty liver, which is softer on palpation and rarely associated with either splenic or renal disease. The prognosis is always grave. When albuminuria sets in, the case always terminates fatally.

Treatment.—Iodide of potassium and iodide of iron are the remedies to be directed against the constitutional taints with which waxy liver is associated.

Dr. Budd recommends the hydrochlorate of ammonia in five-grain doses. The springs of Karlsbad, Ems and Weilbach are recommended. Moffat would appear the most suitable in this country. In a word, alkaline thermal and thermal sulphureous springs have appeared the most beneficial.

Post-mortem.—The liver in these cases is usually found large, smooth and firm; its cut surface pale and glistening. On applying a solution of iodine a deep red tint, characteristic of amyloid matter, is produced; the subsequent addition of sulphuric acid changes this to a deep violet or, more rarely, blue tint.

The spleen is usually enlarged, its consistence firm, and presenting on section the appearance known as "sago spleen," in which the little bodies like grains of sago become blue when treated with iodine and sulphuric acid. The kidneys are also commonly enlarged, and in a state of granular degeneration.

Hydatids of the liver is a disease giving rise to tumor, but very rarely met with in children. The symptoms are few or none until the tumor attains considerable size, and especially when the hydatid cysts are buried in the substance of the gland. More often, however, a rounded bulging tumor is discernible in the right or left hypochondrium or in the epigastrium; such tumors are smooth, globular, and elastic, sometimes imparting a distinct sense of fluctuation. A vibratory thrill has been described as very characteristic when present; it then indicates a multivesicular cyst. Pain is but rarely present, as also jaundice and ascites. There is no fever, but symptoms referable to the pressure exercised by the growth become manifest as the disease goes on. Such are dyspnœa, palpitation, constipation, and œdema. The cyst may burst in various directions, *e.g.*, into the pleura, peritoneum, intestines, or externally. In such cases pleurisy, pneumonia and expectoration of the hydatids, vomiting or evacuation of the hydatids, occur, according to the direction of the bursting. The most favorable condition is when the tumor bursts into the stomach or intestines. Occasionally hydatids undergo spontaneous cure, and are only found post-mortem, their existence having been unsuspected during life. The spleen often suffers from hydatids contemporaneously with the liver. When the diagnosis is doubtful Récamier recommends puncture with a fine trocar. The fluid withdrawn should be limpid, watery, and free from albumen, and the hooklets or scolices of the echinococci are readily recognizable by the microscope.

Common salt and iodide of potassium have been recommended in this disease. Puncture has frequently been attended with success when the tumor is fixed by adhesions to the abdominal wall. Care must be taken to prevent any of the fluid running into the peritoneal cavity; this is best effected by pressing the abdominal wall against the cyst, and afterwards applying a padded bandage and ensuring complete quiet for twenty-four hours. To determine the existence of adhesions, Dr. Budd recommends that the edge of the tumor and liver margins be marked with ink, and then it should be noted if these boundaries are altered by change of posture, deep inspiration, and especially if the most prominent or presenting portion of the tumor remains a fixed point.

Cancer of the Liver is also extremely rare; as a secondary deposit, I have seen one case in a child seven years of age. The diagnosis would be assisted by the cancerous cachexia, by pain, emaciation, sallowness, and feculent diarrhœa. The tumor is hard and nodular, and tender to the touch, though these conditions are not uniformly present. Jaundice, when it occurs, is persistent. Ascites is commonly present. The spleen is usually unaffected and not enlarged; this is of importance, as a diagnostic from amyloid liver. Dyspepsia is usually an early and constant symptom. Hæmorrhages are common towards the close of the disease. The treatment can, unfortunately, be but palliative.

Cancer of the Kidney is another condition occasionally producing abdominal tumor in the child. The same cancerous cachexia, gastric disorder, diarrhœa, and ascites are present, as in the liver cancer; occasionally there is hæmaturia, which, when present, is a significant symptom. Cancer of the right kidney is usually separated from the liver by a coil of intestine, which aids the diagnosis. Cancer of the left kidney is diagnosed from enlarged spleen by its rounder outline in front, the spleen having a sharp edge, and by the greater extent of the tumor into the lumbar region.

Cancer of the Stomach, equally rare a disease, gives rise to greater dyspepsia, to hæmatemesis, and to pain after every meal. Moreover, the percussion is tympanitic when the stomach is diseased, dull when the liver is affected.

Leucocythæmia.—This condition, which appears due to the presence of an excess of white corpuscles in the blood, is attended with enlargement of the spleen especially, occasionally of the liver, thyroid gland, and supra-renal capsules. It is attended with pallor, great emaciation, and debility, tumid abdomen from splenic enlargement, diarrhœa, and a disposition to hæmorrhage, especially epistaxis, and hæmatemesis, anorexia, nausea, jaundice; œdema and ascites are other concomitant symptoms. The disease is not uncommon in children; it is to be recognized from tuberculosis by the splenic and hepatic enlargement, and the extreme pallor, which, in one child that I attended, was most remarkable; this child was carried off by an intercurrent attack of measles. During the year it was under my observation it suffered from general debility, but no special symptoms, except some increasing feebleness; the child remained fairly nourished, and suffered but little. A very mild attack of measles rapidly proved fatal.

The treatment is unsatisfactory. Tonics are recommended. Steel and cod-liver oil seemed to me to be somewhat beneficial, though I cannot say much improvement took place under their use. Iodide of iron is perhaps the best form. Dr. Broadbent and Dr. Wilson Fox have had cases in adults much benefited by phosphorus. Dr. W. Fox, indeed, "trusts that one more disease has thus been rescued from the list of the almost hopelessly irremediable." $\frac{1}{60}$ to $\frac{1}{100}$ of a grain will suffice two or three times a day (see Formulæ).

17. Diseases of the Kidneys.

Diabetes and Diuresis.—True diabetes is so infinitely rare in childhood that it is unnecessary to consider it in this place, the more as, when it does occur, the treatment would be precisely similar in diet and remedies as in the adult. A simple diuresis occasionally follows a more or less prolonged condition of disordered stomach. In such cases thirst is present and the child wastes, often apparently causelessly. The urine is generally transparent pale yellow, of a specific gravity of 1010 to 1025, containing an excess of urea. The strumous diathesis is commonly found in conjunction with this complaint. When it occurs, attention to the diet (animal, and farinaceous foods and milk being most suitable), and careful regulation of the gastric or intestinal disorder, suffice for its cure. Warm baths and change of air are also valuable adjuncts in this affection, which is by no means common.

Acute Desquamative Nephritis rarely arises idiopathically, but is common as a sequela of scarlet fever; the symptoms of the disease are similar in either case, viz., chilliness and feverishness, restlessness, pains across the loins, vomiting. Dropsy is usually an early symptom; the face is the first part affected; a peculiar puffiness is noticed, then the body generally becomes swollen, and there is occasionally effusion into one or other of the serous cavities, especially the pleura, hence the need of frequent auscultation; the urine becomes scanty, albeit micturition is frequent, its color is dark, and it is highly albuminous; a favorable prognostic is an increase in the quantity of urine, on which the dropsy rapidly subsides. During convalescence large quantities of urine, varying from three to six pints, are often passed in the twenty-four hours. Microscopical examination of the urine discovers crystals of lithate of ammonia, mucus, epithelium-scales, casts of the tubules, and blood-corpuscles.

The treatment must be directed to the proper exercise of the skin functions, and for this the hot-air bath is of great value, and with it diaphoretics should be combined. Cream of tartar and Dover's powder in small doses at bedtime is a useful combination. When the disease is yielding and the anasarca diminishing, mild diuretics, as the citrate of potash with a little spirit of nitre, will be useful. It will be necessary throughout to keep up a sufficient action of the bowels, and for this purpose no medicine is better than compound jalap powder. Dr. Dickinson recommends that the patient should drink large quantities of water to dilute the urine and diminish the quantity of albumen. Dr. Johnson recommends wet cupping over the loins in severe cases. Dry cupping always seems to me, in children especially, a safer and very effectual remedy; especially after scarlatina there is no blood to be lost. It need, perhaps, scarcely be added that blisters are most inadmissible. During convalescence cold must be carefully guarded against, and the diet be well regulated. Tincture of steel and quassia are often useful even at an early stage in the convalescence of this affection.

18. Incontinence of Urine

is occasionally associated with renal diseases, as gravel, excess of lithic acid in the urine, also with general debility, worms, loaded bowels, and masturbation. More often it is caused by an excessive use of liquids or by lying on the back during sleep. If no specific cause be present in the urine, if there be no worms and no bad habits, the best treatment will be to have the child held out once or twice every night at the same hour; to tie a reel of cotton over the spine to prevent the dorsal decubitus; to administer internally tincture of belladonna in small doses. A belladonna plaster may also be put over the sacrum. In very obstinate cases a suppository of belladonna in cacao butter is valuable. Abstinence from, or great moderation in, fluids towards night is, of course, necessary. A child seldom requires scolding for this habit; it is usually beyond its control; but if it be due to sheer idleness and dirtiness, as it sometimes is, proper correction must be resorted to. A combination of belladonna and strychnia is sometimes useful, as—

℞ Liquor Strychniæ.............. ℳj.
   Tinct. Belladonnæ.............. ℳij., v. or x., cautiously increased.
   Inf. Cascarillæ................ ℨij.
Ft. mist.

For a child three years old, may be given three times a day; or, benzoic acid in doses of from one to five grains made into a pill, with extract of liquorice, may be tried, and will sometimes succeed when belladonna has failed.

In some cases, especially where periodicity seems an element in the incontinence, tincture of cantharides will be useful. By the irritation, says Dr. Condie, it produces about the neck of the bladder, the moment the urine begins to flow a degree of strangury occurs, sufficient to awaken the patient. From three to six up to ten or fifteen drops three times a day should be given, according to the age of the child, the dose being increased till strangury is caused. It should be employed only in protracted cases unconnected with any derangement of the digestive organs or a morbid state of the urine. If the strangury be too violent, it may be abated by the use of emollient enemata and mild mucilaginous drinks, camphor, and a warm hip bath, or an anodyne enema. Cubebs has been used by Deiters, a few grains for infants, for older children ℨ ss. three times a day, the cubebs should be continued from three to eight weeks. Chloral and ergot have also been recommended, and might be worth a trial in obstinate cases.

Dysuria, or painful micturition, may occur in both sexes. In boys the common causes are very acid urine, a long prepuce, an inflamed state of the meatus urinarius, stone in the bladder, ascarides. In girls, acid urine, inflammation about the meatus, or small warts or vascular growths thereabouts, and protracted leucorrhœa are the causes. The treatment will of course vary with the cause; when that is abnormally acid urine (often associated with chronic skin affections and the rheumatic diathesis) the pain is generally severe, the urine very high colored, and there may be smartish fever and considerable dyspepsia as concomitants. Diluents and alkalies, especially the citrate of potash, mucilaginous and bland drinks, restricted diet, avoidance of all heating food and stimulants, and an occasional warm bath will cure the affection. The other causes are mostly surgical, and being recognized will require appropriate surgical treatment unnecessary to specify here.

Occasionally in new born infants anuria or total suppression of urine occurs. This may continue twelve or even twenty-four hours without exciting much anxiety. The bowels should be relieved, a warm bath employed, and a few drops of sweet spirits of nitre administered; if the cause be not congenital or dependent on hopeless organic changes in the kidney, these measures will probably relieve it.

Retention of urine dependent on spasmodic closure of the urethra is not very uncommon in childhood. It may be caused by chill, more often by worms, occasionally by inflammation about the neck of the bladder. The treatment will comprise a warm bath, a purge, or better, an enema, and the use of the catheter if required. In the case of inflammatory action about the neck of the bladder, small doses of Dover's powder or extract of henbane and ipecacuanha are appropriate remedies, and the catheter will probably be required.

19. VAGINITIS.

Leucorrhœa is by no means uncommon in girls of all ages; generally the seat of discharge is limited to the vulva; it is seen in scrofulous children, and not unfrequently occurs in the course, or as a sequela of, scarlatina. Dentition and ascarides are fruitful causes of this condition. The discharge is not be distinguished from that of gonorrhœa, but the history of the case, the absence of marks of injury, the amount of swelling, the presence of an unbroken hymen, and the presence or absence of painful micturition at the onset of the attack, will generally decide the true origin. The disease runs often a very tedious course, except when due to evident and removable sources of irritation, *e. g.*, worms, dentition and constipation. Most important in the treatment is absolute cleanliness; the discharge must be cleansed away with warm water, as in ophthalmia, several times a day, a weak astringent lotion being afterwards employed —sulphate of zinc, alum, acetate of lead, or nitrate of silver, according to the severity of the case. It is a good plan to use first one and then another of these lotions. Cold salt-water baths and friction with a rough towel are very useful, so is change of air to the sea-side and sea-bathing. Ferruginous and bitter tonics should be administered; the astringent lotions and tonics should be continued for three or four weeks after apparent cure, as this affection is very liable to return.

20. PROLAPSUS ANI.

This is a very common affection, especially amongst the children of the poor, from their too constant habit of giving purging physic of one sort or another to their children. Another cause is the practice of leaving children to strain upon their stools for twenty minutes or half an hour together. Sometimes it is a result of habitual constipation, sometimes a sequence of the straining caused by severe attacks of dysentery and diarrhœa, which have weakened the muscular coats of the bowel. The treatment will vary somewhat with the cause. If there has been abuse of purgatives, especially of "Steedman's powders" and calomel, jalap, &c., such must be prohibited and purgatives given only in degree absolutely necessary to maintain liquid stools. Syrup or confection of senna (castor oil is objectionable, as it tends to constipation), olive oil, or even cod-liver oil, are often sufficient and quite harmless. The diet should be regulated. Brown bread may be given instead of white, a little gruel allowed daily or some stewed fruit, especially apples. Locally the bowel must be carefully returned, pressed back with a well-oiled finger; cold water should be plentifully applied after each protrusion. Cold salt-water hip-baths, and friction of the abdomen and loins with a rough towel, are also useful. I have seldom found much benefit from astringent injections; still they are frequently recommended, and can do no harm. If attention to the bowels and cold-water sponging are insufficient, the child should be compelled to pass its motions lying on its back; it is a good plan, in obstinate cases, to keep the child in bed, with its feet raised up on a pillow, for a few weeks. Surgical operations, *e.g.*, pinching up a fold of the bowel, or destroying a portion of the relaxed membrane, are occasionally necessary. When the condition is the result of weakening diseases, tonics, ferruginous and bitter, may be usefully employed, and iron or bark baths daily used.

CHAPTER IX.

GENERAL THERAPEUTICAL HINTS AND FORMULARY.

The action of a drug upon children often differs in kind, as well as in degree, from its action in the adult. Drugs found comparatively inert in the diseases of grown people often act with surprising effect and vigor in the young child. Children are more tolerant of some drugs than adults, and less tolerant of others; calomel and opium are notable examples. Children are altogether more susceptible, by reason of the greater delicacy of their frames, to the action of powerful agents upon tender organizations; this is a point too much lost sight of by many practitioners. Children tolerate well, and, indeed, require certain classes of remedies which would not be at all suitable for corresponding complaints in the adult; emetics form a good example. Children bear much loss of blood very badly, and general bleeding should I think never be resorted to; leeches are often of value, and the application of a few at the right time, and in the right place, may do much good in the relief of suffering and in the general constitutional effect produced.

Blisters should be resorted to as seldom as possible; every physician must have seen the baneful effects of blisters on the sick child—the extreme anguish they cause, the excitability they set up, the albuminuria they bring on, the intractable sores they so often leave behind. The Emplastrum Lyttæ should be exploded, and when actual blistering is necessary (as for the relief of serous effusion, &c.) blistering fluid should be used, and the sore encouraged to heal as rapidly as possible. Behind the ears and on the vertex are the best points for head blistering; the nape of the neck is the refinement of cruelty, as the child can rest in no way. Physicians whose faith in the action of medicine on the adult has been severely shaken are often astonished and pleased in witnessing the effect of the same medicines in the child. There are just two or three rules for prescribing which are thoroughly practical and should always be borne in mind.

1. It is not good practice to give a number of powerful remedies at the same time. Some practitioners seem to endeavor to exhaust the Pharmacopœia on one sheet of paper, and frequently end with prescribing an incompatible or inert compound. Such a drug as arsenic, for instance, does not require half a dozen others to aid its operation; if good result, the good cannot be credited to the arsenic amidst the crowd; if harm, it is unfair to blame the most potent agent, when, perhaps, some of the incapables have been deranging its action.

2. Remedies of moderate bulk and of no extreme nauseousness should be selected for children. It is the best practice to make their medicine palatable, lest the struggle of giving it counterbalance the benefit produced. An excellent practitioner of my acquaintance has an unfortunate

fancy for assafœtida in the nervous diseases of little ones, with what spasmodic results many an unhappy nursery can testify.

3. Narcotics and irritants must be avoided as far as possible, or when given their effects must be most carefully watched.

4. With regard to dose, no rules can be laid down without exceptions, but the following table of Gaubius is a general and valuable guide:

Ages.	Proportional quantities.	Doses.
Suppose the ordinary adult dose to be 1 drachm, then		
A child under 6 months......	will require $\frac{1}{15}$	2 grains.
" 1 year	" $\frac{1}{12}$	5 "
" 2 years	" $\frac{1}{8}$	7·5 "
" 3 years	" $\frac{1}{6}$	10 "
" 4 years	" $\frac{1}{4}$	15 "
" 7 years	" $\frac{1}{3}$	20 "
" 14 years	" $\frac{1}{2}$	½ drachm.
" 16 years	" $\frac{2}{3}$	2 scruples.
" 21 years	full dose	1 drachm.

FORMULARY.

N.B.—In the following formulæ the doses are such as to be generally suitable for a child three years old. When a lower and a higher dose are given, *e.g.*, Potass. Citrat., gr. v.—xv., the lower dose is intended for a child about three years, the higher for a child about nine or ten years old.

For convenience it will be useful to group together a few of the more important drugs under the ordinary headings of purgatives, emetics, tonics, &c. Each group referred to will be found in the index at the end of this section. The section makes no pretension to completeness, but is inserted simply as a guide to the combination and selection of remedies generally serviceable in the disorders of childhood.

1. Blood-restorers.

Iron—Manganese—Cod-Liver Oil.

Ferrum (Iron).
a. *Without astringency* (pure blood-restorers).
Ferri et Ammoniæ Citras. *Dose*, gr. j.—v. May be given in tincture of orange and water, or in Inf. Gent. co., or in water.
— et Quiniæ Citras. *Dose*, gr. j.—v. May be given in water and sweetened.
— Iodidum. Almost always given as syrup. *Dose*, ℞xx.—ʒj. (Invaluable in scrofula, &c.)
— Phosphas. Chiefly as the compound syrup, with the phosphates of lime, soda, and potash. *Dose*, ℞xx.—ʒj. Parrish's Chemical Food.

I have used this remedy in thousands of cases with the best result; it is tolerated when other preparations of iron are not, and has never disagreed with any little patient to whom I have administered it. Its pleasant taste and pretty color are not without their advantages for sick children. Its value in rickets, in scrofula, and general debility, cannot be over-estimated; I occasionally add a small quantity of quinine or cod-liver oil to each dose.

The syrup of the lacto-phosphate of iron has recently been introduced. I have occasionally employed it, but chiefly in adults. Easton's syrup, containing iron, quinine, and strychnia, is another excellent tonic.

Ferrum Redactum. Dose, gr. ¼—gr. j. A very convenient form of administering iron as powder by reason of its small bulk and tastelessness; is free from astringency, and a good remedy in anæmia, chorea, &c. The British Pharmacopœia has lozenges of this preparation, the dose of which is one or two twice a day.

1. ℞ Ferri Redacti .... gr. xxx.
   Pepsinæ Porci .... gr. xxx.
   Zinci Phosphatis .... gr. xv.
   Glycerini, q. s. ut fiat massa.
   Divid. in pil. xxx. Dose j. or ij. In anæmia, chlorosis, &c. (*Tanner*).

Ferrum Tartaratum, used as wine. Dose, ♏xx.— ʒj. Slightly astringent, not incompatible with alkalies; very largely used in children's disorders.

β. *With astringency.*

Tinct. Ferri Perchlor. Dose, ♏ij.—♏xv. or xx. Very valuable, when combined with Quassia or Calumba, for worms and also as a general tonic; it should be sweetened, the taste being extremely astringent. It cannot be prescribed with alkalies.

ADDITIONAL FORMULÆ.

Iron in effervescence.

2. ℞ Ferri Citrat. .... gr. iij.—vj.
   Acidi Citrici. .... gr. v.—x.
   Aquæ. .... ʒss.—j.
   Cum. Pot. Bicarb. .... gr. v.—x.
   Syrup. Aurant. .... ʒss.—j.
   Aquæ. .... ʒss.—j.
   Ft. mist. cap. dum effervescend. ter die.

With Quinine.

3. ℞ Quinæ Disulph. .... gr. ½.
   Ferri Sulphat. .... gr. ½.
   Acid. Sulph. dil. .... ♏v.
   With or without Magn. Sulph. .... gr. x.—xv.
   Sp. Chloroform. .... ♏j.—iij.
   Aquæ .... ʒss.
   Bis vel ter die.

As Electuary.

4. ℞ Ferri Peroxid. Hydrat. (vel Sesquioxidi) .... ʒij.
   Confect. Aurant.,
   Theriaceæ .... āā ʒj.
   Dose.—Half a teaspoonful or a teaspoonful.

Manganese, unastringent tonic, in large doses stated to be cholagogue.

Manganesii et Ferri Carbonatis cum Saccharo. Dose, gr. iij.—v. Not much used.

Ol. Morrhuæ, a sheet anchor of medicine in scrofula, tuberculosis and debility of all sorts, glandular enlargements, rickets, &c. The secret of giving cod-liver oil successfully is *not too give too much*, and to give it at the right time. Small quantities are best to begin with, a few drops for a very young child, ℨ ss.— ℨ j. for older ones, in orange wine, or a little weak nitro-muriatic acid in water, well sweetened. It should be given so as not to clash with meals, or soon after a meal; if before, it spoils the appetite. Bedtime is a good time when it causes sickness; the child lying down immediately afterwards, it is usually well retained. When it causes diarrhœa, and often in rickets, I give it with equal parts of lime water. A little iodide or phosphate of iron may be dissolved in it, or a little phosphorus when the administration of that drug is desirable. As an external application to many obstinate forms of eczema capitis and other cutaneous diseases, I have found it extremely valuable.

If necessary, it may be made into an ointment, as—

5. ℞ Ol. Morrhuæ... ℨ ss.
   Liquoris Potassæ... ℨ ss.
   Adipis, q. s. Ft. unguent. (*Dr. Neligan*).

When cod-liver oil cannot be tolerated, glycerine and cocoa-nut oil are the best substitutes. They should be given in doses of ℨ j.— ℨ ij. two or three times a day. I have tried the Dugong oil, but do not think that it possesses any special merit, nor yet the cod-liver oil emulsions, jellies, &c.; I much prefer the plain oil. Some bear the light brown kinds well, others prefer the pale. The finest sample of cod-liver oil I have seen came from Messrs. Southall's, of Birmingham. Burgundy or claret make good vehicles for cod-liver oil, or it may be given sandwich fashion in a little brandy and water, that is, pouring a little brandy and water at the bottom of the glass, then floating the oil, wetting the side of the glass with brandy and water, and finally pouring a little *rather* stronger over the top of the oil, will make it slip down tastelessly. Ice in the oil also renders it nearly tasteless. If the oil be thick from cold weather it should be warmed and made clear before administration. As a rule children get to *like* it without artificial means of any kind; I am therefore merely supplying hints for possible difficulties.

2. Antacids.

Salts of Potash, Soda, Magnesia, Calcium, Lithium.

Potassa.

Potassæ Acetas renders urine alkaline; it is diuretic, and laxative. *Dose*, gr. iij.—x.— ℨ ss. as a laxative. Acetate of potash has been recently recommended by M. Labat in large doses in croup.

Potassæ Bicarb. renders urine strongly alkaline; it is antacid and diuretic; much used in rheumatism. *Dose*, gr. iij.—xv. Rather more as a diuretic.

— *Citras* renders urine slightly alkaline; it is refrigerant, saline, mildly laxative, and diuretic. *Dose*, gr. v.—xx.

— *Tartras* renders urine alkaline; it is diuretic, purgative, refrigerant. *Dose*, gr. x.— ℨ ss. Purgative, ℨ ss.— ℨ ij.

Liquor Potassæ does not render urine alkaline, but alkalinizes the blood and renders fibrine less plastic, hence its use in serous inflammations, especially those attended with deposit of fibrinous matter. It is also used in rheumatism and cutaneous diseases. *Dose*, ♏iij.—xv. or xx.

FORMULÆ.

6. ℞ Liq. Potassæ .... ♏iij.—xv.
    Syrupi Simp. .... ʒss.
    Inf. Serpentariæ .... ʒj.—iv.
In lithic acid diathesis. Very useful in chronic rheumatism.

7. ℞ Pot. Bicarb. .... gr. iij.—xv.
    Inf. Gent. co. vel Inf. Calumbæ .... ʒj.—iv.
In dyspepsia with acidity.

8. ℞ Pot. Bicarb. .... gr. iij.—x.
    Tinct. Hyoscyam. .... ♏iij.—x.
    Inf. Pareiræ vel Inf. Buchu .... ʒij.—iv.
In acidity and turbid urine. A good diuretic.

9. ℞ Pot. Citratis. .... gr. v.—ʒss.
    Vin. Ipecac. .... ♏v.—xx.
    Syrup. Scillæ .... ♏xx.—ʒj.
    Aquæ vel Dec. Senegæ .... ʒij.—ʒss.
In bronchitis and febrile cough.

Soda.
Soda Tartarata, renders urine alkaline. *Dose*, gr. xx.—ʒij. A mild refrigerant laxative of great value in many febrile affections, seldom disagreeing with the stomach. A capital medicine for children.

Sodæ Bicarb., a valuable antacid, much used in dyspepsia and heartburn; combined with rhubarb and some carminative (*e.g.*, ginger in Gregory's powder) is a good laxative for infants. *Dose*, gr. iij.—xv. or xx. It is less used than the corresponding potash salt as an antilithic, the urates of soda being less soluble than the urates of potash.

Sodæ Phosphas (tasteless purging salts). *Dose*, ʒss.—ʒij. A valuable purgative for children, usually given in broth or milk; in smaller doses diuretic.

FORMULÆ.

10. ℞ Sodæ Tartaratæ .... gr. xx.—ʒj.
    Pot. Nitrat. .... gr. j.—v.
    Syr. Zingiberis. .... ♏xv.—ʒss.
    Aq. Menth. Pip. .... ʒij.—iv.
A good refrigerant in measles and other febrile affections.

11. ℞ Sodæ Bicarb. .... gr. v.—xv.
    Ammon. Sesquicarb. .... gr. j.—iv.
    Tinct. Gent. co. .... ♏v.—xv.
    Inf. Calumbæ .... ʒij.—iv.
For dyspepsia and acidity of stomach, &c.

12. ℞ Sodæ Bicarb. .... gr. v.—xv.
    Inf. Rhei conc. .... ʒss.—ij.
    Inf. Gent. co. .... ʒss.—ij.
Tonic aperient.

Magnesia and *Magnesia Levis*, antacids and largely used laxatives for children. The light variety is stated to act more quickly. *Dose*, gr. v.— ℨ ss. The new Liquor Magnes. Carb. is an imitation of Dinneford's Fluid Magnesia, which has long enjoyed a great reputation. *Dose*, ʒ ij.— ℨ j.

FORMULÆ.

13. ℞ Magnesiæ.. gr. iij.—v.
    Syr. Rosæ... ʒ j.
A laxative for the youngest infants.

14. ℞ Magnesiæ.. gr. v.—x.
    P. Rhei... gr. iij.—x.
    P. Cinnam. co....................................... gr. j.—iij.
A useful aperient.

Calcium.
Creta Præp. (prepared chalk). *Dose*, gr. iij.— ℨ ss. Antacid and astringent; much used in diarrhœa.
Mist. Cretæ. *Dose*, ʒ j.— ℨ ss. Generally combined with Catechu, Kino, Cinnamon, or Opium.
P. Cretæ Arom. *Dose*, gr. v.—xx.
— — *cum Opio.* *Dose*, gr. ij.—x. (Opium 1 in 40).

FORMULÆ.

15. ℞ Pot. Bromid.. gr. j.—iij.
    Mist. Cretæ... ʒ j.—ij.
    Syrupi.. q. s.
In the irritable stomach of young children, accompanied with vomiting of sour and curdled character.

16. ℞ Cretæ Preparatæ................................... ʒ iss.
    Acaciæ Pulv.,
    Sacchari Albi....................................... ā ʒ j.
    Tinct. Opii... ♏ x.
    Aquæ.. ℨ iij.
    Ft. mist.
A teaspoonful every hour. In diarrhœa (*Dewees*).

Calcis Phosphas. *Dose*, gr. j.—iij. It may be mixed with the food, and is highly recommended by Dr. Beneke in rickets, diarrhœa, ulcerations, and excoriations of the skin and mucous membranes, and in general wasting of children.

17. ℞ Pulv. Calcis Phosphatis............................ gr. xv.
    Bismuthi Nitratis................................... gr. xv.
    Pulv. Sacch. Albi................................... ʒ j.
    Divid. in chartulas 5. Sumat j intercibos nocte maneque. In chronic diarrhœa and wasting (*Trousseau and Reveil*).

The syrups of the hypophosphites of lime, soda, and iron combined in various forms are now largely used in the foregoing complaints, as also in phthisis, &c.

3. ASTRINGENTS.

The following remedies may be grouped under this head:
Aluminum.
Plumbum.
Quercus (chiefly external use, as injections and lotions).
Galla.
Krameria.
Rosa.
Tormentilla (seldom used).
Granati Radicis Cortex (chiefly as anthelmintic).
Hæmatoxylum (decoction, a good remedy, but stains linen).
Kino (the compound powder contains opium 1 in 20. *Dose* according to opium).
Catechu.
Bael (Extract. Liquid. Belæ Fruct. *Dose*, ℥v.— ʒ ss. in dysentery).
Matico (has a tonic action on the urinary passages).
Gummi rubrum (a new remedy). *Dose*, gr. j.—vj. Syrupus Gummi Rub. *Dose*, ℥xx.— ʒ j. Much used in diarrhœa and dysentery.
Hamamelin. (*Dose* of the concentrated tincture one to four drops). In diarrhœa, dysentery, and hæmorrhages.

FORMULÆ.

18. ℞ Aluminis .... gr. iij.—xv.
    Acidi Sulph. dil. .... ℥ij.—x.
    Syrupi .... q. s.
    Inf. Rosæ Acid. .... ʒj.—iv.
In chronic diarrhœa and passive hæmorrhage. To be given every three or four hours.

19. ℞ Aluminis .... gr. xxv.
    Extracti Conii .... gr. xij.
    Syrupi Rhœados .... ʒ ij.
    Aquæ Anethi .... ℥ iij.
    Ft. mist.
In the second stage of pertussis (*Golding Bird*). A dessertspoonful every four or six hours.

20. ℞ Plumbi Acetatis .... gr. j.
    Pulveris Opii .... gr. ½.
    Pulv. Glycyrrhiz. .... gr. iij.
    Ft. pulvis.
In the hæmorrhage of typhoid fever. To be given every six hours for a child five years old.

Acidum Gallicum. Dose, gr. j.—v. Dissolved in hot water, and well sweetened and allowed to cool. Useful in chronic diarrhœa, excessive sweating, and in hæmorrhages; also in bronchitis with excessive expectoration.

21. ℞ Acidi Tannici .... gr. v.
    Acidi Nitrici dil. .... ℥vj.
    Inf. Gentian. co. .... ʒ ij.
    Ft. mist.
To restrain excessive secretion and in hæmorrhage. Every three or four hours for a child ten years old.

Glycerine of tannin, as an application, is highly praised by Dr. Ringer in ozæna and other offensive nasal discharges, in chronic otorrhœa and vaginitis, in chronic inflammation of the throat, in relaxed throat and uvula causing irritable cough, and in eczema behind the ears of children.

Combinations of the infusions or tinctures of Kino, Catechu, and Krameria, with the mineral acids, chalk or opium, are the ordinary remedies for diarrhœa. One or two formulæ will suffice.

22. ℞ Tinct. Kino.. ♏x.
        Syrup. Papav. Alb... ♏x.
        Tinct. Catechu.. ♏v.
        Aquæ Cinnamom... ʒij.
    Every three or four hours, as an astringent.

23. ℞ Extract. Rhataniæ... ʒss.
        Confect. Ros. Gall.. ʒiv.
        Syrup. Papav. Alb... ʒij.
        Pulv. Catechu... gr. xv.
    Ft. electuarium.
    In diarrhœa, &c. (*Trousseau*). Dose—a teaspoonful.

4. ACIDS.

Acidum Sulphuricum (dil.). Tonic, refrigerant, and astringent. *Dose*, ♏ij.—v. or x.

— Nitricum (dil.). Tonic, refrigerant, alterative. *Dose*, ♏ij.—v. or x.

— Nitro-hydrochloricum (dil.). Tonic, hepatic, alterative. *Dose*, ♏j.—v.

— Citricum } Refrigerants, diminishing thirst.
— Tartaricum } *Dose*, gr. ij.—x.

— Phosphoricum (dil.). Tonic, refrigerant; used in rickets. *Dose*, ♏ij.—v.

— Sulphurosum. Antiseptic; destroys vegetable growths. *Dose*, ♏v.—xx.

— Hydrochloricum (dil.). Refrigerant, antiseptic, and tonic. *Dose*, ♏ij.—v.

— Carbolicum. Antiseptic, astringent; very valuable in chronic offensive diarrhœa. *Dose*, gr. ¼—j., largely diluted and sweetened.

FORMULÆ.

24. ℞ Acid. Sulph. Aromatic..................................... ♏ij.
        Aquæ.. ʒij.
    In diarrhœa; for a child five years old.

25. ℞ Acidi Phosphorici dil..................................... ♏iij.
        Acidi Hydrochlor. dil..................................... ♏iij.
        Inf. Calumbæ.. ʒiv.
    In mucous and phosphatic urine.

26. ℞ Acidi Nitrici dil... ♏ij.—vj.
        Syrupi.. q. s.
        Aquæ vel Decoct. Hordei................................... ʒiv.—ʒj.
    In typhoid and other fevers.

27. ℞ Acidi Nitrici dil................................. ♏ij.—x.
       Inf. Chirettæ, Inf. Gent. co., vel Decoct. Cinchonæ... ʒiv.—ʒj.
Stomachic tonic.

28. ℞ Acidi Nitrohydrochlor. dil......................... ♏ij.—v.
       Syrupi Sarsæ.................................. ʒj.
       Aquæ ....ad. ʒiv.
In syphilitic cachexia.

29. ℞ Acidi Nitrohydrochlor dil......................... ♏ij.—v.
       Ext. Tarax. Liquid............................. ♏x.—ʒss.
       Inf. Cascarillæ .... ʒij.—iv.
Hepatic alterative, and tonic.

5. ALTERATIVES.

Iodine (chiefly as Syr. Ferri Iodidi and Pot. Iodid.).
Bromine (chiefly as Pot. Bromid. and Ammon. Bromid.).
Chlorine }
Sulphur   } (in their compounds).
Arsenic.
Mercury.
Potass. Chlor.
Ammon. Hydrochlor.
Dulcamara.
Sarsa.
Hemidesmus Indicus.
Irisin. *Dose* of the concentrated tincture two to eight drops; stated to be specially useful in scrofula, and syphilis.
Phytolacin. *Dose* of tincture one to four drops.
Corydalin. *Dose* of the tincture one to four drops.

FORMULÆ.

Iodide of Potassium.

30. ℞ Pot. Iod.................................... gr.½.—vj.
       Sp. Am. Arom................................. ♏j.—v.
       Syr. Sarsæ................................... ♏x.—ʒj.
       Aquæ....................................... ʒij.—ʒss. or ʒj.
In syphilitic skin diseases and cachexia, &c.

31. ℞ Pot. Iod.................................... gr.½—v.
       Tinct. Hyoscyam............................. ♏ij.—x.
       Inf. Serpentariæ............................. ʒij.—iv.
Chronic rheumatism, syphilitic affections, &c.

32. ℞ Pot. Iod.................................... gr.½—v.
       Sp. Am. Arom................................. ♏j.—v.
       Magn. Sulph.................................. gr. x.—xx.
       Aq. Camph................................... ʒij.—iv.
Chronic pleurisy with effusion.

33. ℞ Pot. Iod.................................... gr.½—v.
       Ferri Ammon. Citrat........................... gr. ij.—x.
       Syr. Sarsæ................................... ♏xv.—ʒss.
       Aquæ.. ʒij.—iv.
In debility when the action of Iodine is desirable.

34. ℞ Syr. Ferri Iodidi............................. ♏xx.—ʒj.
       Dec. Cinchonæ................................ ʒij.—iv.
Scrofulous affections, strumous glands, &c.

35. ℞ Pot. Iodidi.. gr. ½—v.
    Glycerini... ʒss.—j.
    Syr. Aurantii.. ʒss.—j.
    Ex Aquâ. Bis vel ter quotidie.
In tuberculosis and the early stages of phthisis.

36. ℞ Pot. Iodidi.. gr. ½—v.
    Ol. Morrhuæ... ʒss.—ij.
    Aq. Calcis... ʒss.—ij.
    Bis vel ter die.
In phthisis, &c.

Bromide of Potassium.

37. ℞ Pot. Bromid.. gr. j.—v.
    Syr. Papav. Alb..................................... ♏x.
    Vin. Ipecac.. ♏v.—x.
    Dec. Senegæ... ʒij.
In pertussis. To be given every two or three hours, the quantity of the Bromide increased gradually. For a child one year old.

38. ℞ Potassii Bromidi.................................... gr. iij.—x.
    Syrup. Simp... ♏xv.
    Aquæ... ʒij.
To be given two or three times a day. In splenic and hepatic enlargements, convulsions, insomnia, and nervous excitability.

39. ℞ Potass. Bromidi..................................... gr. iij.—x.
    Tinct. Lobeliæ Inflat............................... ♏v.—x.
    Syr. Rhœados.. ʒss.
    Aquæ... ʒij.
Every four hours. In laryngismus stridulus.

Bromide of Ammonium.

40. ℞ Ammonii Bromidi.................................... gr. ½—v.
    Tinct. Valerianæ Ammoniat......................... ♏x.
    Aquæ... ʒij.
In spasmodic affections and pertussis.

41. ℞ Ammonii Bromidi.................................... gr. ij.
    Tinct. Hyoscyam.................................... ♏x.
    Syrup. Simpl.. ♏xv.
    Aquæ... ʒij.
A draught at bedtime. In night-terrors and sleeplessness.

Arsenic. Extensively employed in chronic skin diseases, in chorea, dyspepsia, &c. It should be given directly after or even with meals; and ophthalmia, smarting in the conjunctivæ, nausea, colicky pains, and diarrhœa, are indications for the reduction of the dose. Most dermatologists agree that Arsenic is not beneficial during the acute stages of skin diseases. Arsenic is one of the drugs that children seem to tolerate better than adults, even in large doses.

42. ℞ Liquoris Arsenicalis................... ♏½—iij., or even v.
    Tinct. Hyoscyam....................... ♏v.
    Inf. Calumbæ........................... ʒij.
To be taken three times a day immediately after meals. In chronic skin affections, especially psoriasis, eczema, lepra, and pemphigus. The Henbane may be omitted, but it sometimes causes the Arsenic to agree better when the stomach is irritable. The effect is to be closely watched, and the dose may be cautiously increased.

43.	℞	Liquoris Arsenicalis.............................	*♏ ½—iij.
		Quinæ Disulph....	gr. ¼—ij.
		Acidi Sulphurici dil...........................	♏ iij.
		Syr. Zingiberis................................	♏ xx.
		Aquæ..	ʒ ij.

Three times a day. In chorea, atonic dyspepsia, &c.

44.	℞	Liquoris Arsenici Hydrochlorici...............	♏ ½—ij.
		Aquæ Camph...................................	ʒ ij.

This preparation of the Ph. B. is about three times *stronger* than that of the Ph. Lond. It is stated to agree better with the stomach than the Liquor Arsenicalis. Half a drop is sufficient for a child five years old.

Liquor Sodæ Arsenatis. Dose, ♏ ½—j.
Ferri Arsenias. Dose, gr. $\frac{1}{60}$—$\frac{1}{30}$.
Iodide of Arsenic. Dose, gr. $\frac{1}{100}$—$\frac{1}{60}$.

The Iodide of Arsenic is formed extemporaneously when the Liquor Arsenicalis and Liquor Potass. Iodid. comp. are ordered together.

Mercury.

45.	℞	Hydrarg. c. Cretâ................................	gr. iij.
		Pulv. Rhei,	
		Pulv. Jalap. co................................	āā. gr. iv.

An ordinary aperient when the liver is sluggish.

46.	℞	Hydrarg. c. Cretâ................................	gr. ½—ij.
		Pulv. Cinnam. co.............................	gr. j.—iij.

Two or three times a day to produce the constitutional effects of Mercury in young children.

47.	℞	Calomelanos....................................	gr. j.
		P. Sacch. Alb...................................	gr. iij.

A convenient and effectual purgative for infants; it can be placed on the back of the tongue, and is readily swallowed.

48.	℞	Calomelanos....................................	gr. j.
		P. Pot. Nitrat...................................	gr. j.
		P. Sacch. Alb...................................	gr. ij.

In inflammation, to be repeated every three or four hours.

49.	℞	Calomelanos....................................	gr. ij.
		Pulv. Jalap. co................................	gr. v.
		Pulv. Ipecac...................................	gr. ½.

In the early stages of bronchitis and pneumonia.

50.	℞	Liq. Hydrarg. Perchlor........................	♏ ij.—x.
		Tinct. Cinchon. co............................	♏ v.—xx.
		Aquæ..	ʒ ij.—iv.

In syphilitic cachexia.

Hydrarg. Iodid. Viride. Dose, gr. $\frac{1}{8}$—$\frac{1}{4}$.
— *Rubri.* Dose, gr. $\frac{1}{20}$—$\frac{1}{15}$.

51.	℞	Hydrarg. Iodid. Rubri.........................	gr. j.
		Potassii Iodidi.................................	ʒ ij.
		Liquoris Arsenicalis...........................	ʒ iss.
		Tinct. Lavand. co.............................	ʒ ij.
		Spirit. Chloroformi...........................	ʒ iv.
		Aquæ...	ad. ʒ xij.

Dose.— ʒ ss.— ʒ j. three times a day after meals. In psoriasis and some inveterate squamous, tubercular and ulcerous affections of the skin. (*Dr Tanner.*)

52. ℞ Hydrarg. Perchlor............................ gr. ij.
    Ætheris....................................... ℥ j., solve et adde.
    Olei Jecoris Aselli........................... ℥ vj.
℥ j. contains 1-24th grain of the Perchloride. (*Bumstead.*)

Chlorate of Potash.

53. ℞ Pot. Chlorat................................. v.—x.
    Syrup.. q. s.
    Aquæ... ℥ ij.—iv.
In typhoid and other fevers, severe thrush, stomatitis, and affections of the mouth and throat.

54. ℞ Pot. Chlorat................................. gr. ij.—x.
    Tinct. Cinchon. comp......................... ℳ v.—xx.
    Syrupi....................................... q. s.
    Aquæ... ℥ j.—iv.
In ulcerative stomatitis, diphtheria, &c.

55. ℞ Pot. Chlorat................................. ℥ j.
    Decoct. Hordei, vel Lactis, Oj. Pro potu.
In fevers.

Ammon. Chloridum (Ammoniæ Hydrochlor.). *Dose*, gr. ij.—v. As an expectorant with stimulant properties; as an alterative in glandular enlargements; by some considered cholagogue.

56. ℞ Ammon. Chlorid.............................. gr. ii.—v.
    Syrup. Hemidesmi............................. ℥ ss.—j.
    Aq. Cinnamom................................. ℥ ij.—iv.
    Fiat mist. quartis horis.
In adynamic febrile affections, subacute bronchitis, &c.

57. ℞ Ammon. Chlorid.............................. gr. ij.—iv.
    Liq. Ammon. Acetat........................... ℳ xx.— ℥ ss.
    Syrupi Limonis............................... q. s.
    Aq. Camph.................................... ℥ iv.
In typhoid.

58. ℞ Ammon. Chlorid.............................. gr. ij.—iv.
    Oxymel Scillæ................................ ℳ xx.— ℥ ss.
    Syrup. Tolutan............................... ℳ xx.
    Decoct. Althææ............................... ℥ iv.
In catarrhal affections.

Sarsa, } Their infusions and decoctions are useful adjuncts to
Dulcamara. } Iodide of Potassium, &c.

Hemidesmi Radix (chiefly used as syrup in kidney affections.) It is a slight diuretic. *Dose*, ℥ ss.—j.

59. ℞ Ammonii Iodidi............................. gr. j.—iij.
    Syr. Hemidesmi............................... ℥ ss.
    Inf. Cascarillæ.............................. ℥ ij.—iv.
In strumous glandular enlargements, &c. Two or three times a day.

60. ℞ Sodii Iodidi................................ gr. iv.
    Syr. Sarsæ................................... ℥ ss.
    Dec. Sarsæ................................... ℥ iv.
In syphilitic cachexia, when Iodide of Potassium does not agree.

6. STIMULANTS TO THE SPINAL CORD.

Strychnia.
Arnica.
Rhus Toxicodendron.

Strychnia.
Liquor (℥ij. contains gr. j. of the alkaloid). *Dose*, ♏ss.—ij. or iij. cautiously given.
Tinct. Nucis Vomicæ (strength 1—10). *Dose* ♏j.—v.

61. ℞ Tinct. Nucis Vomicæ............................ ♏j.—v.
        Tinct. Cinchon. co............................. ♏v.—xv.
        Decoct. Cinchon............................... ℥ij.—iv.
In paralysis and some stomach diseases.

62. ℞ Tinct. Nucis Vomicæ............................ ♏j.—v.
        Syr. Ferri Phosph. co.......................... ♏xv.—℥j.
        Aquæ... ℥ij.—iv.
A powerful tonic where the use of Strychnia is indicated.

63. ℞ Liquor Strychniæ............................... ♏ij.
        Tinct. Belladonnæ.............................. ♏v.
        Inf. Cascarillæ................................ ℥ij.
Two or three times a day for a child five years old. . In incontinence of urine.

64. ℞ Liquor Strychniæ............................... ♏ij.
        Tinct. Cinchon. co............................. ♏v.
        Inf. Calumbæ................................... ℥ij.
In intestinal irritation occurring shortly after meals.

65. ℞ Extract. Nucis Vomicæ.......................... gr. iij.
        Fellis Bovini Inspiss.......................... gr. xij.
        Extract. Taraxaci.............................. gr. xx.
        Extract. Gentianæ.............................. gr. xx.
        Divid. in pil. xxiv. Sumat j, bis die.
In habitual constipation for children nine or ten years of age, and young girls shortly before the menstrual period is established.

Arnica, of the tincture. *Dose*, ♏x.—℥ss.

66. ℞ Inf. Arnicæ (from ℥j. of the flowers)............ ℥iv.
        Syrup. Croci................................... ℥vj.
        Æth. Sulph..................................... ♏x.
        Coch. j. larg. Secundis horis.
For a child three years old. In hydrocephalus. (*Dr. Ure.*)

67. ℞ Tinct. Arnicæ.................................. ♏x.
        Tinct. Lavand. co.............................. ♏v.
        Inf. Serpentariæ............................... ℥ss.
Stimulant to the nervous system.

Arnica is also useful in small doses for nocturnal incontinence of urine.

Rhus Toxicodendron. The powdered leaves made into infusion or tincture are highly recommended in paraplegia. The dose of the powdered leaves is gr. ⅛—¼. The action is stated to resemble that of Strychnia. But Rhus also produces diuresis and diaphoresis, on which account it has been used in erysipelas and fevers. It is a drug requiring further investigation. That its action is very powerful cannot be doubted, and it should therefore always be given with caution. Dr. Neligan employed the tincture in doses of from ℥ss.—j. in adults; ♏v.—xv. would therefore be sufficient for a child.

7. SEDATIVES TO THE SPINAL CORD.

Potass. Bromid.
Ammon. Bromid.
? Conium.
Calabar Bean.
Gelsemin.
Scutellarin.

Bromide of Potassium is now given in large doses in epilepsy and other convulsive nervous disorders. In the convulsions of children (in whom the spinal system is peculiarly excitable) it is extremely efficacious. In insomnia, restlessness, laryngismus stridulus, and hooping-cough, it is also largely employed. The dose may be increased even in young children from gr. j.—v. or even x., until the sedative effect is manifested. It is a remedy which I have used with much success in the earliest stages of hydrocephalus.

Conium.

Succus Conii Fol. Dose, ♏v.—xv. or xx. Tinct Conii Fruct. *Dose*, ♏v.—xv. Anodyne and antispasmodic; used in chorea, pertussis, and the irritative cough of bronchitis and bronchial phthisis. There is considerable difference of opinion as to its method of action. The Succus appears to be the most active preparation. I have failed to get satisfactory results with even large doses of the Extract.

68. ℞ Succi Conii.. ♏xx.
     Syr. Papav. Alb... ♏x.
     Mist. Amygdal... ʒij.
In pertussis for a child five years old.

Physostigmatis Faba or *Calabar Bean* has recently been much used in tetanus with undoubted advantage; and in chorea and other affections of the nervous system. The dose of the powdered bean is from gr. ij.—iv. for a child seven years old.

Gelsemin is a new American remedy, of very great efficacy in spasmodic and nervous disorders. *Dose*, gr. $\frac{1}{12}$—$\frac{1}{4}$. Of Keith's concentrated tincture, one drop to five; the dose must not be too large, or "double vision" and other unpleasant symptoms will result. The effects of this remedy in many diseases of the nervous system are most markedly salutary.

Scutellarin and *Gelsemin*, together or separately, are medicines destined to rank highly in the treatment of many cases of nervous disorder. In neuralgia, in over-excitability of the nerves, so common in children, in wakefulness, restlessness, apparently causeless peevishness, these are remedies far more valuable than opium, henbane, or any others I know of. They do good without doing harm. Convulsions, chorea, and I venture to think, many of the more serious affections of the brain and spinal cord in children, will be found to be benefited by these unquestionably powerful medicines. I recommend caution in their use, which must at present be partly tentative, but I am satisfied that those who use them will not be disappointed, and I trust that we shall soon be in possession of fuller facts regarding their use. I think it questionable if they are truly sedatives to the spinal cord. I have elsewhere* called attention to the great value of gelsemin in facial neuralgia.

---

* "Lancet," 1877.

8 ANTISPASMODICS.

Assafœtida.
Sagapenum.
Galbanum.
Valerian.
Rue.
Camphor.
Sumbul (of the tincture. Dose, ♏ij.—x).
Succini Oleum.
Castor.
Musk.

FORMULÆ.

69. ℞ Tinct. Assafœtidæ... ʒss.
        Syrup. Rhœados... ʒj.
Dose.— ʒj. every hour. In flatulent colic.

70. ℞ Tinct. Assafœtidæ... ♏v.
        Oxymel Scillæ.. ♏x.
        Tinct. Opii.. ½♏.
        Syrup. Rhœados.. ʒj.
        Aquæ ad... ʒj.
To be taken frequently in pertussis. For a child two years old.

71. ℞ Tinct. Valerian. Ammoniat.................................... ♏v.—x.
        Tinct. Camph. co.. ♏v.
        Sp. Chloroformi... ♏iij.
        Aquæ Anethi... ʒij.
In laryngismus and other spasmodic affections.

9. STIMULANTS.

Alcohol, ⎫
Ether,   ⎬ These are more especially stimulants to the brain.
Chloroform,⎭

Ammonia,   ⎫ These are considered to be stimulants to the gan-
Phosphorus,⎭ glionic system.

Turpentine,
Resins,
Myrrh,
Lavender,
Rosemary,
Mentha,   ⎬ May be grouped as vascular stimulants.
Cinnamon,
Cajeput,
Nutmeg,
Clove,
Ginger,
Cardamomum, ⎭

Tar,
Creosote,
Petroleum,   ⎬ Antiseptics.
Carbolic Acid,
Liquor Sodæ Chloratæ,⎭
Salicin and Salicylic Acid.

Alcohol in small quantities increases the secretion of the gastric juice, hence its value in atonic dyspepsia. In too large quantities it destroys the powers of digestion, retards oxidation, and impairs the general health. Healthy children are infinitely better without alcohol in any form; it is not a food, and if habitually taken is less valuable as a medicine when really required in illness; for Alcohol, although a bad food, is a good medicine. Perhaps the tendency of the present day is to overrate its powers, —the effects of the reaction from bleeding and blistering of days gone by. Still, in cases of great exhaustion, in the continued fevers, and in general debility, there is no doubt as to its value. For children, port wine and brandy are, I think, on the whole, the best forms for its administration.

FORMULÆ.

72. ℞ Ammoniæ Sesquicarb.................................. gr. j.—iij.
       Tinct. Zingiberis...................................... ℳv.
       Syrup. Aurant... ʒss.
       Aquæ Cinnamomi..................................... ʒss.
Stimulant in exhaustion, &c.

73. ℞ Aquæ Menth. Pip..................................... ʒiss.
       Sp. Ammon. Arom.................................... ʒss.
       Sp. Ætheris Nitrosi.................................... ℳxij.
       Sp. Lavandulæ co...................................... ʒj.
       Syrup. Simpl.. ʒss.
Dose.—ʒj. every two hours. In receded eruptions and sinking from exhaustion. (*Evanson* and *Maunsell.*)

74. ℞ Ætheris vel. Sp. Ætheris............................... ℳv.
       Sp. Chloroformi...................................... ℳv.
       Spir. Myristicæ....................................... ℳx.
       Inf. Caryophylli...................................... ʒiij.
A powerful diffusible stimulant. For a child five or six years old.

75. ℞ Liquoris Sodæ Chloratæ............................... ʒj.
       Tinct. Cinchon. co.................................... ʒvj.
       Spiritus Vini Gallici................................... ʒxij.
       Aquæ ad.. ʒviij.
       Ft. mistura.
A dessert or tablespoonful in low fever with great prostration. (*Dr. Tanner.*)

Phosphorus.

76. ℞ Phosphori.. gr. ss.
       Olei Succini... ʒss.
ℳj.—ij. ter die, ex aqua. In paralysis and loss of nerve power.

77. ℞ Phosphori.. gr. j.
       Ol. Morrhuæ... ʒj.
This should be allowed to stand for fourteen days in a dark place, then add
       Olei Caryophylli...................................... ℳv.
Dose.—ℳv. in mist. Amygdal. ter die sumend. In some cases of rickets and phthisis, the dose to be increased with great caution if the stomach bear it well.

It would seem as if the profession had but recently awakened to the value of Phosphorus in medicine. For my own part I think it difficult to over-estimate its value. I leave the above formulæ as representing the forms in which Phosphorus was formerly used by the few who believed in it. Recently, however, not only in neuralgia, where its efficacy is undoubted (and shows the influence of the remedy upon the nervous system), but in many other affections indicating "want of tone," phosphorus has been recognized to be of real service. In epilepsy, hysteria, chorea, tuberculosis and its many manifestations, scrofula, and defective nutrition and debility generally, its employment is gaining ground every day. I have seen that fact sufficiently demonstrated in New Zealand. The forms usually employed in Australia and New Zealand are those of Kirby & Co., and the various combinations with nux vomica, reduced iron, and quinine, present special advantages in special cases. That with nux vomica and iron is, for anæmia and chlorosis, perhaps the very best. I can give but one or two of these admirable preparations taken from Dr. Kirby's work —"Selected Remedies."

78. ℞ Phosphori Pur. .... gr. $\frac{1}{33}$—$\frac{1}{50}$.
        Ferri Redacti. .... gr. ij.
        Sol. Nucis Vom. .... gr. $\frac{1}{4}$.
     Ft. pil.
One two or three times a day, with or directly after food.

79. ℞ Phosphori Pur. .... gr. $\frac{1}{50}$.
        Quiniæ Sulph. .... gr. j.
     Ft. pil.
One pill three times a day (child from seven to ten years).

Phosphorus can be given by itself or in other combinations, and I must refer those who wish further information to Dr. Kirby's little book. I have had one or two combinations of my own prepared by Messrs. Kirby & Co., and they have always succeeded in carrying out my wishes (as expressed through Messrs. Sharland & Co. of this place) to my satisfaction. Those nearer home can have no difficulty in obtaining any experimental combinations they may desire.

Terebinth. Ol. Dose, ♏ij.—v. As an anthelmintic ʒj.—ij. or more with castor oil. Turpentine is reputed stimulant, diaphoretic, diuretic and astringent.

80. ℞ Ol. Terebinth. .... ʒss.
        Ol. Limonis. .... ♏iv.
        Syr. Simp. .... ℥ss.
        Aq. Cinnam. .... ℥j.
     Ft. mist.
Dose—ʒj., quart. q. hor. In diarrhœa and flatulence. (*Evanson* and *Maunsell*.)

81. ℞ Inf. Rosæ. .... ʒij.
        Magn. Sulph. .... gr. x.
        Mannæ. .... gr. x.
        Ol. Terebinth. .... ♏iij.
To be given every four hours. In hæmatemesis, &c.

Tar is employed externally in the form of ointment in chronic eczema, and internally also in skin diseases and chronic catarrhal affections. It is given internally in the form of capsules. *Dose,* gr. iij.—vj

Creosote has been employed in chronic bronchitis, in phthisis to restrain secretion, in neuralgia, in skin affections, and in chronic vomiting. Its taste is extremely unpleasant. *Dose,* ♏$\frac{1}{4}$—$\frac{1}{2}$ in flavored mucilage.

Carbolic Acid is now extensively used both externally as a lotion or glycerine, and also internally. It is an admirable antiseptic. I have employed it considerably in some forms of chronic feculent diarrhœa with excellent results. It has been given in typhoid fever. As a gargle, one or two grains should go to an ounce of water. *Dose* of the glycerine, ♏j.—iv. freely diluted. *Dose* of the solid acid (Calvert's), gr. ¼—¾ in water well sweetened.

Charcoal and Permanganate of Potash are also much used as antiseptics, but they are destitute of stimulating properties.

Salicin and *Salicylic Acid* have recently been employed in acute rheumatism with excellent results. The dose of Salicin for a child seven years old or so, would be five to ten grains every two, three, or four hours, according to the severity of the case. I have referred to this matter in the body of the work.

Salicylic Acid is largely used as a disinfectant. Dr. Ogilvie Will extols it as an ointment in eczema, especially of the head and face, in children, in eczema rubrum, and in E. impetiginodes. He uses one drachm of acid to one ounce of lard.

10. Sedatives to the Brain and General Sedatives.

Opium, { Morphia, Codeia, }
Rheas,
Lactuca,
Cannabis Indica,
Belladonna.
Atropia.
Stramonium.
Hyoscyamus.

} These are regarded as more particularly sedatives to the brain.

FORMULÆ.

Opium is one of the drugs which children do not tolerate so well as adults, and which, therefore, always requires caution in its administration.

82. ℞ Tinct. Opii.. ♏j.
       Syrup. Croci... ℥ss.
       Aquæ.. ℥ss.
    Dose— ʒj. every two or three hours for a child six months old.

83. ℞ Tinct. Camph. co..................................... ♏xv.
       Syrup. Rhœados...................................... ʒij.
       Aquæ Camph.....................................ad. ℥j.
    Dose— ʒj. For the very youngest infants when Opium is required.

Syrupus Codeiæ. *Dose*, ʒj. Has been used in pertussis.

The *Liquor Morphiæ Acetat.* (Ph. B.) is useful in pertussis. The dose may be a minim to commence with, gradually increased and given with sufficient frequency to control the cough. Combined with Belladonna it is very efficacious.

84. ℞ Liquor. Morph. Acetatis............................ ♏ viij.
       Extract. Belladonnæ............................... gr. ij.
       Oxymel. Scillæ.................................... ʒ ij.
       Aquæ Camph................................... ad. ʒ j.

Dose—ʒ j. for a child one year old with pertussis. The Belladonna may be gradually increased up to one or two grains at a dose for children five years old, but it is always best to commence with a small dose.

85. ℞ Tinct. Camph. co................................. ♏ xx.
       Vin. Ipecac....................................... ♏ xx.
       Syrup. Tolutan.................................... ʒ ij.
       Mucilag. Acaciæ.............................. ad. ʒ j.

Dose—ʒ ij. every four hours for a child four years of age, in severe coughs.

The *Solution of Bimeconate of Morphia* of similar strength I have used very frequently, and I consider it the best preparation of Morphia we have for children or adults.

Belladonna has been much used in pertussis. It is, perhaps, the most valuable drug we possess against that disease. It requires to be given cautiously. I have repeatedly seen the characteristic eruption thrown out, with dryness of the fauces and dilated pupils, from extremely small doses. At the same time cases will be met with requiring large quantities before any effect is manifested. It is desirable to use a fresh extract.

86. ℞ Extract. Belladonnæ............................ gr. ½—j.
       Potass. Bromid................................... gr. j.—v.
       Syrup. Papav..................................... ♏ xv.
       Aquæ.. ʒ ij.

In pertussis to be given every two or three hours, night and day, until the paroxysms are reduced in number and severity.

87. ℞ Extract. Belladonnæ............................ gr. ½.
       Syrup. Simpl..................................... ʒ j.
       Aquæ Destill..................................... ʒ ij.

Dose—For a child one year old, ʒ j. every hour, in pertussis. (*Bouchut.*)

88. ℞ Tinct. Belladonnæ............................... ♏ ij.—v.
       Liquor Strychniæ................................. ♏ j.—ij.
       Syrup. Simpl..................................... ʒ ss.
       Inf. Canellæ..................................... ʒ ij.

In incontinence of urine.

89. ℞ Atropiæ.. gr. ¾.
       Sacch. Albi...................................... ʒ iiss.
       Misceantur optime.

Dose—A grain or a grain and a half two or three times a day for a child five years old with pertussis. (*Bouchardat.*)

Externally *Belladonna* is serviceable in relieving pain. Plasters of the drug are commonly employed in cardiac affections and painful diseases of the chest and elsewhere.

Stramonium is used chiefly in asthma and neuralgia. It is not much employed for children.

Lactucarium, a drug of rather doubtful efficacy. It is supposed to add diuretic to its narcotic properties. *Dose*, gr. j.—iij. or v.

90. ℞ Lactucarii .... ʒss.
       Decoct. Lichen. Islandic. .... ʒij.
       Mucilaginis. .... ʒss.
       Syrupi. .... ʒj.
       Ft. Mist.
Dose—ʒij., frequently in spasmodic cough, &c. (*Brera*.)

Hyoscyamus I have found to be a remedy of the greatest service in children's diseases. It is a safe and efficient sedative without the deleterious effects of Opium.

91. ℞ Tinct. Hyoscyam. .... ♏v.—x or xv.
       Liq. Ammon. Acet. .... ♏x.— ʒss.
       Vin. Ipecac. .... ♏v.—xv.
       Aq. Camph. .... ʒij.—iv.
In some forms of bronchitis, &c.

92. ℞ Tinct. Hyoscyam. .... ♏v.
       Syrup Papav. .... ♏v.
       Aquæ Anethi. .... ʒj.
An anodyne for a young infant.

93. ℞ Extract. Belladonnæ. .... gr. ij.
       Tinct. Hyoscyam. .... ♏xx.
       Syrupi Simp. .... ʒss.
       Aquæ. .... ad. ʒiss.
Dose— ʒij., for a child five years old. In incontinence of urine.

Cannabis Indica.

94. ℞ Tinct. Cannab. Indic. .... ♏j.—v.
       Mucilaginis. .... ʒj.
       Triturat. et adde
       Aquæ. .... ʒij.— ʒiv.
This drug is a powerful anodyne and antispasmodic. It has lately been extensively used in insomnia, neuralgia, and spasmodic affections. It is stated to cause none of the unpleasant after-effects of opium. It should be cautiously given. In children pertussis, chorea, irritative cough, laryngismus stridulus, are the diseases in which it is most serviceable. Mr. Squire directs that it should be ordered as in the text, or the resin will be precipitated by the water.

11. Vascular and Heart Sedatives.

Antimony.
Bismuth.
Colchicum.
Hydrocyanic Acid.
Digitalis.
Aconite.
Lobelia.
Actæa Racemosa.
Veratri Viridis Radix.
Tabacum.
Aqua Laurocerasi (*Dose*, ♏¼—j.; seldom used).

Hydrate of Chloral.—Used a good deal as a soporific. M. Ferrand recommends five grains in simple syrup for a child four or five years old with pertussis, but if given many successive nights unpleasant symptoms may appear.

FORMULÆ.

Antimony.

95. ℞ Pulv. Jacobi Ver.................................. gr. j.—iij.
    Calomelanos.. gr. j.
    Sacch. Albi.. gr. v.
    Ft. pulv.
Dose—Every four hours. In acute inflammations.

96. ℞ Vin. Antimonial.................................. ℥ ss.
    Vin. Ipecac.. ♏ xv.
    Tinct. Camph. co................................... ♏ xx.
    Mucilag. Acaciæ,
    Syrupi Scillæ...................................... āā ʒ iv.
Dose—ʒ j., every two or three hours. In bronchitis, &c., for a child two years old.

97. ℞ Antim. Tartarat................................. gr. j.
    Potass. Nitrat..................................... ʒ j.
    Mist. Amygdal...................................... ℥ iij.
Dose—ʒ j., every two hours. In pneumonia, bronchitis, &c.

Bismuth (subnitrate and carbonate).

98. ℞ Bismuthi Albi (subnit.),
    Magn. Carb... āā gr. xvi.
    Acidi Hydrocyan. dil............................... ♏ v.
    Aquæ... ℥ iv.
Dose—ʒ ij., for a child three years old. In flatulent gastrodynia and gastralgia.

99. ℞ Bismuthi Carb.................................... gr. j.—ij.
    Magn. Carb... gr. iij.
    Tinct. Hyoscyam.................................... ♏ v.
    Inf. Rhei.. ʒ ij.
In atony and irritability of the stomach.

Prussic Acid (Acid. Hydrocyanici dil. *Dose*, ♏ ⅛—j., always to be cautiously given).

100. ℞ Acidi Hydrocyan. dil............................ ♏ ij.
    Tinct. Hyoscyam.................................... ♏ xx.
    Syrup. Aurant...................................... ʒ ss.
    Mist. Amygdal...................................... ℥ ij.
    Ft. mistura.
Dose—ʒ ij. may be given frequently in pertussis, laryngismus stridulus, croupy cough, &c., for a child five years old. ʒ j. for a child two years old.

101. ℞ Acidi Hydrocyan. dil............................ ♏ x.
    Syrup. Papav. Alb.................................. ʒ iij.
    Aq. Flor. Aurant................................... ad ℥ vj.
Dose—ʒ j., capiend. secund. vel tert. horis. In spasmodic cough for a child three years old, the dose may be cautiously increased.

Colchicum is used chiefly in gout and rheumatism; not much employed in children's diseases.

102. Vin. Colchici.................................... ʒ iij.
    Spirit. Ætheris Nitrosi............................ ʒ ij.
    Pot. Acetat.. ʒ ij.
    Aquæ... ad ℥ iv.
Dose—ʒ j., every four hours. In scarlatina, when there is suppression of urine, high fever, and delirium. (*Dr. Bennett.*)

Digitalis is considered to be a cumulative drug. It is a drug requiring caution, as it certainly acts sometimes out of all proportion to the dose in which it is given. In cardiac affections it is particularly valuable for palpitation and tumultuous action caused by hypertrophy. When dropsy is present it is especially indicated; its value as a diuretic then also comes into play. It is one of our best remedies in scarlatinal dropsy.

103. ℞ Tinct. Digitalis... ℳiij.
         Tinct. Hyoscyam... ℳv.
         Syr. Aurantii... ʒss.
         Aq. Camp... ʒiv.
For a child five years old may be given every six hours.

104. ℞ Infus. Digitalis... ʒss.
         Pot. Acetatis... gr. v.
         Spirit. Juniperi co... ℳx.
         Decoct. Scoparii...ad ʒiv.
In anasarca every four or six hours. Child five years old.

105. ℞ Pulv. Digitalis.. gr. vj.
         Hydrardg. Subchlor.. gr. xij.
         P. Sacch. Alb.. gr. xviij.
         Misce et divid. in pulveres xij.
One powder every six hours in hydrocephalus. (*Dr. Merriman.*)

106. ℞ P. Digitalis Fol.,
         Potassæ Nitratis... āā ʒss.
         P. Sacchari Albi... ʒiiss.
         Misce et divid. in chartulas xl.
One powder thrice daily, in inflammations of lungs and heart, and in dropsies. (*Trousseau.*)

107. ℞ Infusi Digitalis.. ʒviiss.
         Potassæ Nitratis.. ʒij.
         Acidi Hydrocyanici dil..................................... ℳxiv.
         Syrup. Aurantii... ʒij.
Dose—ʒj. every two hours. In hypertrophy of the heart, with excessive action. Child five years old. (*Dr. Copland.*)

Aconite, one of the most powerful sedatives which we possess. It is anodyne, depressant, and antiphlogistic. In the early stages of acute inflammations it is of value; the pulse is lowered, the circulation is rendered less rapid, excitement calmed, and moisture appears upon the skin. I have recently employed Aconite to a rather large extent at the instance of a homœopathic friend; and while I have failed altogether in obtaining results from such a preparation as the third homœopathic decimal, either in myself or in others, yet with ¼- or ½- drop doses of the tincture of the Ph. B. the results are such as above described. I may just mention that Aconite is of undoubted value in acute rheumatism, in fact it has long been used in this disease. The circumstance appears to me to destroy the homœopathic notion of its use. The inflammation of rheumatism differs materially from ordinary acute inflammations; to mention only one point—it exhibits no tendency to suppuration. According to homœopathic law one remedy cannot be homœopathic to two different conditions. It is clear, therefore, that Aconite cannot be homœopathic to both kinds of inflammation. I mention this matter because I have met practitioners who appear to have an objection to employ Aconite lest it should be thought homœopathic treatment. Supposing that this objection were valid against the use of any means calculated to relieve suffering, the above consideration appears to me completely to remove it. Aconite is further useful in pertussis, neuralgia, excited action of the heart in hypertrophy, pericarditis, &c.

I have not altered the above paragraph, which appeared in the first edition of this book nearly eight years ago. But I have now to add my experience of these years, which is still entirely in favor of Aconite, employed in small and frequent doses, in almost all acute inflammations. I have referred very frequently throughout the work to this remedy in the special diseases referred to. I feel, therefore, that it is only necessary in this place to confirm generally what has elsewhere been particularly insisted upon.

108. ℞ Tinct. Aconiti (Ph. B.) .... ♏ $\frac{1}{4}$—$\frac{1}{2}$—j.
Syrup. Croci .... ♏ x.
Aquæ Camph .... ℥ ij.
To be given every hour. In broncho-pneumonia, parotitis, and the early stages of acute inflammation generally.

109. ℞ Tinct. Aconiti (Dub.) .... ♏ iij.
Mist. Camph .... ℥ ss.
Ft. haust. 4tis horis.
In acute rheumatism at fourteen or fifteen years of age. (*Dr. Neligan.*)

Lobelia. The action of Lobelia is sedative, diaphoretic, and expectorant. It is used in asthma, catarrhs, croup, pertussis, and bronchitis.
Dose of the tincture, ♏ ij.—v. or x.
Dose of the ethereal tincture, ♏ ij.—v. or x.

110. ℞ Tinct. Lobel. Æthereæ .... ♏ iij.
Syrup. Hemidesmi .... ℥ ss.
Decoct. Malvæ .... ℥ ij.
For a child three years old. In paroxysmal coughs the dose may be cautiously increased.

111. ℞ Tinct. Lobel. Æthereæ .... ♏ iv.
Succi Conii .... ♏ x.
Syrup. Croci .... ♏ xx.
Mist. Amygdal .... ℥ ij.
To be given every three or four hours. In pertussis. Child five years old.

Green Hellebore. *Dose* of the powdered root, gr. $\frac{1}{2}$—j, or ij. *Dose* of the tincture (Ph. B.), $\frac{1}{2}$—iij. or v. This drug is now extensively used in America, like Aconite, in acute inflammations, especially pleurisy, pneumonia, and peritonitis. It is also given in the spasmodic affections of children. As it occasionally produces unpleasant symptoms it must be given with caution.

Actæa Racemosa. This drug is employed chiefly in chorea and rheumatism. *Dose* of the tincture, ♏ iij.—x. or xv.

12. NERVINE TONICS.

Silver.
Zinc.
Copper.
Cinchona.
Quina.
Beberiæ Sulphas.
Salicine.

Argent. Nitrat., valuable especially in spasmodic nervous diseases, eclampsia, chorea, &c. *Dose*, gr. $\frac{1}{24}$—$\frac{1}{12}$ or $\frac{1}{6}$, in the form of pill made with crumb of bread or any of the following extracts, which may appear most suitable.

> Fell. Bovin. Inspissat.
> Ext. Gentianæ.
> — Conii.
> — Hyoscyam.
> — Humuli.

112. ℞ Argent. Nit. .... gr. $\frac{1}{16}$—$\frac{1}{6}$.
   Aquæ Destillat. .... ℥ ij.
   Syr. Simpl. .... ℥ v.
  Ft. mist.
Coch. min. j., vel ij., quart. q. hor. In obstinate diarrhœa. (*Trousseau.*)

Zinc.

113. ℞ Zinci Sulphatis. .... gr. $\frac{1}{4}$.
   Tinct. Cinchon. co. .... ♏ x.
   Inf. Calumbæ. .... ℥ ij.
Tonic.

114. ℞ Zinci Sulphatis. .... gr. $\frac{1}{4}$.
   Acidi Sulph. dil. .... ♏ ij.
   Syr. Aurant. .... ℥ ss.
   Inf. Aurant. .... ℥ ij.
Tonic. (*Dr. Druitt.*)

115. ℞ Zinci Valerianatis. .... gr. $\frac{1}{4}$—$\frac{1}{2}$.
   Syr. Hemidesmi. .... ℥ ss.
   Aq. Flor. Aurant. .... ℥ ij.— ℥ iv.
In chorea, three or four times a day.

Copper.

116. ℞ Cupri Sulphatis. .... gr. j.
   Syrup. Papav. .... ℥ j.
   Aquæ Anisi. .... ℥ iij.
Dose— ℥ j., every three or four hours. In pertussis. (*Chavasse.*)

117. ℞ Cupri Sulphatis. .... gr. j.
   Ext. Hyoscyam. .... gr. ij.
   Ext. Gentianæ. .... gr. v.
  Ft. pil. viij.
Dose—One every four hours. In chorea, chronic diarrhœa, &c.

Cinchona.
Pale bark is best suited for irritable stomachs.
Yellow bark is a more powerful tonic, but apt to disagree.
Red bark, containing both Quinine and Cinchonine, is the most potent tonic of the three.
 Doses—Tinct. Cinchon. Flavæ. ♏ x.— ℥ ss.
   Ext. Liquid. Cinchon. Flavæ. ♏ v.—x.
   Tinct. Cinchon. comp. (of pale bark). ♏ x.— ℥ ss.
   Extract. Cinchon. Rub. (Ph. Lond.) gr. j.—iij.

118.	℞	Ammon. Sesquicarb.................................	gr. iij.
		Tinct. Cinchon. co................................	♏ x.
		Dec. Cinchonæ....................................	ʒ ij.

Every three or four hours. In debility, &c.

119.	℞	Acidi Nitrici dil..................................	♏ iij.
		Tinct. Cinchon. co................................	♏ x.
		Syrup. Aurant....................................	ʒ j.
		Decoct. Cinchonæ................................	ʒ iv.

In convalescence taken two or three times a day.

Quina. A tonic in constant use, and more potent than any with which we are acquainted. Small doses as a rule answer best with children. Nor will the stomach always bear Quinine; it is very often advisable to commence a course of tonic treatment, particularly after very exhausting diseases, not with Quinine, but with Gentian, Chiretta, Calumba, or some other of the minor tonics, and as the strength and appetite begin to improve, to introduce Cinchona or Quinine gradually.

120.	℞	Quinæ Disulph....................................	gr. ¼.—j.
		Acidi Sulph., dil.................................	♏ ij.—v.
		Magnes. Sulphat.................................	gr. v.—x.
		Aquæ..	ʒ ij.—iv.

A common form of administration; the Epsom Salts correct the constipating effects of Quinine.

121.	℞	Quinæ Disulph....................................	gr. j.
		Tinct. Valer. Ammon..............................	♏ x.
		Aq. Camph.......................................	ʒ iv.

In chorea, for a child seven years old.

122.	℞	Quinæ Disulph....................................	gr. ij.
		Acid. Sulph. Arom................................	♏ xvj.
		Aquæ Destill.....................................	℥ iss.
		Syrupi Caryophylli...............................	℥ ss.
		Misce.	

Dose—ʒ j.—ij. For very young children. (*Dr. Joy.*)

123.	℞	Ferri et Quinæ Citratis............................	gr. xl.
		Tinct. Aurant....................................	ʒ ij.
		Syr. Aurant......................................	℥ j.
		Aquæ ad...	℥ vj.

Dose—ʒ j.—ij. A very efficient form of tonic, and rather palatable.

124.	℞	Quinæ Valerianatis...............................	gr. x.
		Liq. Taraxaci....................................	ʒ vj.
		Tinct. Sumbuli...................................	ʒ ij.
		Inf. Cascarillæ...................................	℥ v.
		Misce.	

Dose—ʒ j.—iv. In neuralgic and spasmodic nervous affections. (*Dr. Neligan.*)

Beberiæ Sulphas. Dose, gr. ¼—j. Tonic and antiperiodic. May be given in water of bitter infusions.

Salicine. Dose, gr. ½—j. or ij. Tonic and stomachic.

These two substances are used as substitutes for Quinine when that drug cannot be taken or cannot be procured. Salicine is readily borne by the stomach, and is a drug of some value. It has recently come into prominence in acute rheumatism, and for its antiseptic properties. (See pp. 25 and 178).

13. Stomachics.

Calumba (an excellent stomachic).
Cusparia (aromatic, useful in dysentery).
Quassia (very valuable in threadworms).
Simaruba (used in dysentery, very bitter).
Gentian (tonic and stomachic).
Chiretta (as Gentian, a finer bitter).
Cascarilla (an aromatic stomachic).
Canella (aromatic, but bitter).
Cortex Winteri (a warm aromatic).
Salicis Cortex (tonic and antiperiodic—see Salicine).
Lupulus (tonic, diuretic, narcotic. Lupulina—*Dose*, gr. ½—j.). Hops are sometimes used as a pillow to procure sleep.
Aurant. Cortex, }
Limon. Cortex, } (aromatic tonics).
Anthemis (aromatic tonic; in large doses emetic).
Artemisia Absinthium (extremely bitter, anthelmintic).
Pepsine.

FORMULÆ.

125. ℞ Tinct. Lupuli............ ♏x.—xx.
     Acid. Nit. dil............ ♏ij.—v.
     Inf. Cascarillæ.......... ʒij.—iv.
In chronic dysentery, &c.

126. ℞ Tinct. Opii............. ♏j.
     Acidi Nit. dil............ ♏iv.
     Inf. Simarubæ............ ʒj.
Dose—ʒj., in milk or barley water every three or four hours. In chronic diarrhœa.

127. ℞ Tinct. Cardam. co........ ♏v.
     Syr. Aurantii............ ʒss.
     Inf. Cuspariæ............ ʒij.
In flatulent indigestion.

128. ℞ Sodæ Bicarb. Exsicc...... gr. xxiv.
     Extract Taraxaci......... ʒss.
     Syrup. Aurant............ ʒij.
     Inf. Calumbæ............ad ʒij.
     Misce.
Dose—ʒij. Tonic and alterative. (*Dr. Hillier.*)

129. ℞ Acidi Sulph. dil......... ♏xvj.
     Tinct. Aurantii.......... ʒj.
     Syrupi................... ʒj.
     Inf. Aurantii............ ʒj.
     Aq. Cinnamomi........... ʒij.
ʒj. three times a day for a child one year old. In vomiting from weak and irritable stomach. (*Dr. West.*)

Pepsine.
Pepsina Porci is the best preparation. It is valuable in cases of dyspepsia from atony of the stomach, with deficient secretion of gastric juice. *Dose*, gr. ij.—iv. Pepsine Wine (Morson's) is an agreeable form of giving the drug. *Dose*, ʒss. in water. Corvisart's Pulv. Pepsinæ et Amyli or "Poudre Noutrimentive" is required in larger doses, gr. v.—x. before meals. Pepsine has also been prepared in the form of lozenge.

14. EMETICS.

Ipecacuanna.
Tartar Emetic.
Mustard.
Sulphate of Zinc.
Sulphate of Copper.
Ammoniæ Sesquicarb.
Baptisin (gr. ½, a new remedy of American origin); of the concentrated tincture five to ten drops. A valuable antiseptic in typhoidal conditions, putrid sore throat, &c.; in larger doses, emetic.
Alum.

FORMULÆ.

130. ℞ Pulv. Ipecac... gr. ½—j.
        Sacch. Alb... gr. iij.
To be repeated every quarter of an hour till sickness results. Safe for young infants.

131. ℞ Vin. Ipecacuanhæ,
        Syrupus croci................................... āā ʒ j.
Every quarter of an hour till vomiting occurs. The wine is not a good form as an emetic. I have given two ounces of it and failed to cause sickness, even with the help of warm water, tickling the throat, &c. The spirit in the wine restrains the action, and the effect of the ipecacuanha is often spent upon the bowels, causing purgation.

132. ℞ Vin. Ipecacuan................................... ʒ ss.—j.
        Vin. Antimonial.................................. ℳ x.—xx.
        Syrupi... q. s.
Ad emesem' rather depressing, but a quick emetic.

133. ℞ Antim. Pot. Tart................................. gr. ij.
        Oxym. Scillæ....................................... ʒ j.
        Aquæ... ʒ j.
ʒ ij. or ʒ iij. every quarter of an hour in the early stage of croup for a child three or four years old.

134. ℞ Aluminis... ʒ i.
        Syrup. Violæ....................................... ʒ iii.
Dose— ʒ ss. To be repeated if necessary. In croup, &c.

Mustard is a capital emetic, always at hand and useful in emergencies; ʒ j. or ʒ ij. in warm water freely diluted, rapidly produces vomiting without depression.

135. ℞ Cupri Sulphatis................................... gr. ½.
        Syrup. Violæ....................................... ʒ j.
In a little barley water, repeated often till sickness occurs. For a child one year old.

Amongst emetics must now be included Apomorphia, a substance recently discovered by Messrs. Matthiessen and Wright. In a communication read before the Royal Society in June, 1869, these gentlemen state that

$\frac{1}{10}$th of a grain of the hydrochlorate, when subcutaneously injected, has been found by Dr. Gee to excite vomiting in from four to ten minutes. A quarter of a grain taken by the mouth produces a similar effect. Dr. Gee regards the Hydrochlorate of Apomorphia as "a non-irritant emetic and powerful anti-stimulant." One-fortieth of a grain of this substance hypodermically injected produced sickness within a few minutes, in two little patients of mine at the Victoria Hospital. The children were ten and eleven years of age respectively, both suffering from chorea, in which disease this drug, thus employed, appears sometimes efficacious, although the method of its operation remains to be explained.

15. LAXATIVES.

Cassia (Confection. *Dose*, ʒj.—ij.).
Prunum.
Tamarind.
Manna.
Viola.
Ficus.
Sulphur.
Magnesia.
Bran.
Treacle.
Mustard Seeds.

FORMULÆ.

136. ℞ Mannæ Optimæ.. ʒij.
    Aq. Anethi.. ʒj.
Dose—ʒj. *pro re natâ.* For young infants.

137. ℞ Mannæ Opt.. ʒij.
    Syrup. Rosæ.. ʒj.
Dose—ʒj.

138. ℞ Magnes. Carb... gr. xx.
    Mannæ .... ʒij.
    Tinct. Rhei co.. ʒj.
    Syrup. Rosæ ....ad ℥iss.
Dose—ʒj.—ij.

139. ℞ Sulphur. Sublim... ℥j.
    Theriacæ .... ℥ij.
Dose—ʒj. A well-known nursery recipe.

140. ℞ Magnes. Calcin... gr. xl.
    Pulv. Rhei... gr. xx.
    Pulv. Cinnamomi.. gr. x.
    Misce.
Dose—gr. iij.—vj for infants. (*Evanson and Maunsell.*)

16. PURGATIVES.

Purgatives—
    Castor Oil.
    Aloes.
    Rhubarb.
    Jalap.
    Senna.

Drastic Purgatives—
    Scammony.
    Colocynth.
    White Hellebore.
    Croton Oil.
    Buckthorn.
    Gamboge.

Hydragogue Purgatives—
    Elaterium.
    Bitartrate of Potash.

Saline Purgatives—
    Sulphate of Potash.
    Sulphate of Soda.
    Sulphate of Magnesia
    Tartrate of Potash.
    Soda Tartarata.
    Phosphate of Soda.

Mineral Waters—
    Cheltenham.
    Epsom.
    Leamington.
    Püllna.
    Seidlitz.
    Carlsbad.
    Friedrichshall, &c., &c.

Cholagogue Purgatives—
    Calomel.
    Colchicum.
    Leptandrin. (*Dose* of the concentrated tincture, five to ten drops.)
    Podophyllin.

Cholagogue Purgatives (continued)—
    ? Taraxacum.
    Hydrastin. (*Dose* of concentrated tincture, one to four drops, of the powder, quarter to half a grain.)

Anthelmintics—
  Kusso.
  Mucuna.
  Santonin.
  Spigelia.
  Pulvis Stanni.
  Staphisagria.
  Filix Mas.
  Turpentine.
  Granati Radicis Cortex.
  Kamala.

(For the use and doses of these, see the Chapter on Worms.)

FORMULÆ.

141. ℞ Pulv. Rhei... gr. iij.
       Pulv. Scammon. co .... gr. v.
       Pulv. Jalap. co .... gr. v.
An effectual purgative for a child three or four years old.

142. ℞ Calomelanos... gr. ij.
       P. Scammon. co .... gr. x.
A drastic purge when worms are suspected, &c.

143. ℞ Podophyllin.. gr. $\frac{1}{6}$—$\frac{1}{4}$.
       Leptandrin... gr. $\frac{1}{5}$.
       P. Jalap. co... gr. v.
Hepatic purgative.

144. ℞ Tinct. Rhei co.. ʒj.
       Syrupi Sennæ... ʒij.
       Decoct. Taraxaci ....ad ʒj.
Dose— ʒj.—ij., *pro re natâ*.

145. ℞ Magnes. Sulphatis................................. gr. x.—xx.
       Ætheris Chlorici..................................... ♏iij.
       Inf. Gent. co.. ʒij.—iv.
Tonic and aperient.

146. ℞ Ol. Ricini,
       Ol. Terebinth,
       Mist. Acaciæ,
       Aquæ Menth. pip..... ....āā ʒij.— ʒss.
       Ft. haustus.
In tapeworm. For children twelve or fourteen years old. (*Dr. Hooper.*)

147. ℞ Decocti Cort. Rad. Granati........................... ʒxj.
       Syrupi Zingiberis.................................... ʒj.
Dose—A wine-glassful three times a day. In tapeworm when the bowels have been cleared by Castor Oil.

148. ℞ Pulv. Elaterii.. gr. $\frac{1}{8}$.
       Pulv. Scammon. co................................... gr. v.
       Pot. Bitart.. ʒss.
       Ft. pulvis.
A powerful hydragogue purge. Child ten years old.

149. ℞ Tinct. Rhei co .... ʒj.
     Syrupi Sennæ............................... ʒj.
     Syrupi Rhamni.............................. ʒj.
     Aquæ................................... ad ʒj.
A rather brisk purge. For a child ten years old.

150. ℞ Potassæ Sulphatis........................... gr. xii.
     Inf. Rhei.................................. ʒvss.
     Tinct. Aurantii............................. ʒss.
     Aquæ Carui................................. ʒij.
ʒss. for a dose for a child three years old. (*Dr. West.*)

151. ℞ Potass. Bitart............................... ʒij.
     Extract. Glycyrrhizæ........................ gr. xx.
     Decoct. Aloes co............................ ʒvj.
     Aquæ Menth. Pip........................ ad ʒij.
ʒij. *pro re natâ.*

152. ℞ Magnes. Sulphatis........................... ʒij.
     Potassæ Sulphatis........................... ʒss.
     Potassæ Nitratis............................ gr. xxiv.
     Syrupi Limonum............................. ʒij.
     Aquæ.................................. ad ʒij.
ʒij.—iv. Saline aperient. (*Dr. Hillier.*)

153. ℞ Sodæ Phosphatis............................ ʒj.
     Syrupi Limonum............................ ʒss.
     Decoct. Hordei.............................. ʒvj.
     Ft. mist.
Dose—Two tablespoonfuls. An agreeable aperient.

154. ℞ Ricini Olei................................. ʒj.
     Ovi Vitelli semissem. Tere simul et adde
     Aquæ Florum Aurant.,
     Syrup. Simpl........................... āā ʒj.
     Aquæ...................................... ʒvj.
A pleasant way of giving Castor Oil. *Dose*— ʒj., or more. (*Trousseau.*)

155. ℞ Infusi Sennæ co............................. ʒxv.
     Potassæ Tart................................ ʒij.
     Extr. Glycyrrhiz............................ gr. v.
     Tinct. Card co.............................. ʒj.
     Spir. Ammon. Arom......................... ♏xij.
     Fiat mistura.
Dose— ʒij.—iv. An efficient purgative. (*Dr. Underwood.*)

17. DIURETICS.

Pareira
Buchu.
Chimaphila.
Uva Ursi.
Potassæ Nitras.
Juniper.
Scoparius.
Sp. Ætheris Nitrosi (*Dose* ♏v.—xv. or xx.).
Cubebs.

Piper Longum and Piper Nigrum.
Copaiva.
Cantharis.
Many Soda salts are diuretic; Potash salts are so even more decidedly. Also the Benzoate of Ammonia in doses of two or ten grains, especially in dropsies, chronic Bronchitis, &c.

FORMULÆ.

156. ℞ Acidi Nitrici dil. .... ♏ iij.—v.
     Tinct. Hyoscyam. .... ♏ v.—x.
     Decoct. Pareiræ. .... ʒ ij.—iv.
Three times a day, in chronic cystitis.

157. ℞ Pot. Bicarb .... gr. v.
     Tinct. Hyoscyam. .... ♏ v.
     Inf. Buchu .... ʒ ij.
In irritable bladder and acid urine.

158. ℞ Pot. Nitratis .... gr. iij.
     Spir. Ætheris Nit. .... ♏ v.
     Syrup. Croci .... ʒ ss.
     Aquæ .... ʒ iv.
     Ft. mist., quartis horis sumend.
Diuretic and febrifuge.

159. ℞ Sp. Juniperi .... ♏ v.
     Decoct. Chimaphilæ .... ʒ ij.
     Sp. Ætheris Nitrosi .... ♏ v.
In dropsies, &c.

160. ℞ Succi Scoparii .... ♏ x.
     Decoct. Uvæ Ursi .... ʒ ij.
In mucous urine, &c.

161. ℞ Potassii Iodidi .... gr. viij.
     Potassæ Nitratis .... gr. xxxij
     Extract. Taraxaci .... gr. xl.
     Infusi Digitalis .... ʒ j.
     Syrupi .... ʒ ij.
     Aquæ .... ad ℥ iv.
Dose—A tablespoonful for a child six years old. (*Dr. Hillier.*)

18. DIAPHORETICS.

Liquor Ammon. Acetatis.
— — Citratis.
Serpentaria (a very valuable remedy in low febrile conditions, chronic rheumatism, and general cachexia).
Guaiacum (useful in periosteal affections and some skin diseases; stated to be valuable in Tonsillitis).
Mezereon, } (occur in Dec. Sarsæ co.).
Sassafras, }
Opium (in minute doses).

FORMULÆ.

162. ℞ Tinct. Serpentariæ............................... ℳx.
       Decoct. Mezerei............................... ℥iv.
       Tinct. Guaiaci Ammon......................... ℳv.
       Mucilaginis................................... ℳx.

In chronic rheumatic and syphilitic pains.

163. ℞ Liq. Ammon. Acetatis........................... ℥ss.
       Syrup. Rhœados................................ ℥ss.
       Aquæ Flor. Aurant............................ ℥iv.

A pleasant diaphoretic, may be given every three or four hours.

164. ℞ Liq. Ammon. Acetatis........................... ℥ss.
       Tinct. Camph. co............................. ℳv.
       Sp. Ætheris Nitrosi........................... ℳv.
       Aquæ.................................... ad ℥ss.

A draught at bedtime to produce copious diaphoresis.

165. ℞ Acidi Citrici................................. gr. xv.
       Tinct. Guaiaci............................ ℳx.—xx.
       Potass. Bicarb................................ ℈j.
       Mucilaginis.............................. ad ℥j.

To be taken while effervescing every three or four hours. In tonsillitis.

166. ℞ Ammon. Sesquicarb............................ gr. iij.
       Syrup. Aurant................................. ℥ss.
       Inf. Serpentariæ........................... ℥ij.—iv.

In typhoid conditions when diaphoresis is desirable.

19. EXPECTORANTS.

Expectorants—
    Senega.
    Ammoniacum.
    Scilla.
    Balsam of Peru.
    Balsam of Tolu.
    Storax.
    Benzoin.
    Ipecacuanha.
    Antimony.
    Asclepin (*Dose*, gr. ¼—j. So-called "Pleurisy Root," a new remedy). Of the concentrated tincture, five to ten drops, recommended even in advanced phthisis as promoting comfort of patient.

Demulcents—
    Acacia.
    Tragacanth.
    Althæa.
    Cetraria, &c.

FORMULÆ.

167. ℞ Tinct. Benzoin. co........................... ʒij.
     Pulv. Tragacanth........................... ʒss.
     Aquæ Cinnam............................... ʒiij.
Dose—ʒij. In chronic bronchitis.

168. ℞ Vin. Ipecac................................ ♏v.—xv.
     Syr. Scillæ................................. ʒss.
     Oxym. Scillæ............................... ♏v.—x.
     Decoct. Senegæ............................. ʒij.—iv.
In bronchitis when expectoration is viscid and difficult.

169. ℞ Decoct. Senegæ............................ ʒiv.
     Vin. Antimonialis.......................... ♏xx.
     Syrup. Althææ.............................. ʒj.
Dose—ʒj., sæpe sumend. An expectorant mixture for croup and bronchitis during the acute stage.

170. ℞ Mist. Ammoniaci........................... ʒvj.
     Sodæ Bicarb................................ ʒss.
     Tinct. Camph. co........................... ʒij.
     Tinct. Hyoscyam............................ ʒj.
     Vin. Ipecac................................ ʒij.
Dose—ʒj., sæpe urgent. tuss. A valuable formula in phthisis, catarrhal cough, &c.

EXTERNAL APPLICATIONS.

20. BATHS.

A Tepid Bath for a child should have a temperature of about 85° Fahr.

A Warm Bath for a child should have a temperature of about 90° Fahr.

A Hot Bath for a child should have a temperature of about 98° Fahr.

When a child evidently dreads the water, an excellent plan is that suggested by Dr. Eustace Smith, viz.—to cover the bath with a blanket, to place the child thereon, and then gently to lower it into the water. By this simple plan much screaming, terror, and unnecessary exhaustion are avoided.

Ice is a most useful agent in the diseases of children, applied to the head in convulsions, fever, meningitis, &c.; sucked, it is grateful in fevers, and valuable in affections of the throat, *e.g.*, diphtheria and tonsillitis, &c. It is also useful to check sickness and hæmorrhage.

Dr. Chapman's spinal ice-bag is recommended in laryngismus stridulus, chorea, eclampsia, and tetanus.

Blanket Bath.—This is useful in producing ready diaphoresis. A blanket is wrung out of hot water and wrapped round the child. Three or four dry blankets are then thrown over, and the child left for half an hour or so. The body should then be rubbed with a soft "fluffy" towel to absorb the moisture thoroughly, and the child should of course remain in bed.

EXTERNAL APPLICATIONS.

The Wet Compress consists simply of a roll of flannel or soft linen dipped in cold water and wrung out, and then applied to the part indicated; over it a piece of waterproof sheeting may be placed, rather larger than the roll.

The Cooled Bath.—The child is immersed in water at 95° F., which in about thirty minutes is cooled to 70° F., or lower, if necessary, by the addition of cold water. A child may, however, be often wrapped in a wet sheet, and a little cold water poured over its head as a readier measure answering a similar object.

Nitro-Muriatic Acid Bath.

171. ℞ Nitric Acid.................................... 1 fluid ounce.
     Hydrochloric Acid............................ 2 fluid ounces.
     Warm Water................................... 10 gallons.

This must be made in a wooden bath, and the child should remain in it about ten minutes. It is used chiefly for hepatic sluggishness.

Sulphur Bath.

172. ℞ Sulphide of Potassium........................... 2 ounces.
     Warm Water.................................... 10 gallons.

Useful in scabies, and in chorea, and other nervous affections.

Salt-water Bath.

173. ℞ Common Bay Salt, or better Tidman's Sea Salt...... 4 ounces.
     Water, Warm or Cold (according to season, &c.)...... 4 gallons.

To be used every morning in tuberculosis, scrofulosis, general debility, rickets, &c., a most useful remedy. The whole body should be rubbed after every bath with a Turkish towel or rough bath gloves, to excite healthy action of the skin.

Mustard Bath.

174. ℞ Powdered Mustard................................. 2 ounces.
     Hot Water .... 4 gallons.

For a foot-bath, as a derivative, occasionally as a stimulant; in conditions of great exhaustion the child is immersed all but its head.

Iron Bath.

175. ℞ Sulphide of Iron................................. ½ ounce.
     Water.. 4 gallons.

For strumous and rickety children. The Ammonio-Citrate of Iron may be used, but it is more expensive. The Steel bath is useful in some diseases of the skin.

Bark Bath.

176. ℞ Half a pound of either of the Cinchona Barks, boiled for half an hour with a pint of water, and strained. The decoction thus made can be added to two gallons of water. Other forms of Bark may be similarly used.

Gelatine Bath.

177. ℞ Gelatine, 4 ounces (Glue is sometimes used as a cheap substitute); Boiling water sufficient to dissolve it, and added to four gallons of water. This is most useful in many cutaneous affections.

Glycerine Bath.

178. ℞ Glycerine.. ½ an ounce.
     Tragacanth..................................... ½ an ounce.

Boil in a pint of water, and add 4 gallons. In skin affections, &c.

Iodine Bath.

179. ℞ Iodine.................................... ½ a drachm.
    Solution of Potash.......................... ¼ an ounce.
    Water...................................... about 7 gallons.
In tubercular and general cachexia, and in skin affections.

Bromine Bath.

180. ℞ Five drops of Bromine, ¼ an ounce of Iodide of Potassium, and about 7 gallons of water.
In syphilitic and scrofulous eruptions.

Valerian Bath.

181. ℞ One drachm of Valerian Root made into infusion, and added to the water of the bath.
In essential convulsions, eclampsia, &c. (*Trousseau.*)

Corrosive Sublimate.

182. ℞ Corrosive Sublimate................................ gr. x.
    Alcohol.. ℥ ij.
    Distilled Water...................................... ℥ j.
To be added to the bath. In syphilitic skin diseases. (*Trousseau.*)

21. COUNTER-IRRITANTS.

Strong Iodine Paint.

183. ℞ Iodinii................................... gr. x.
    Potassii Iodidi........................... gr. v.
    Sp. Vin. Rect............................. q. s. ut fiat pigment.
To be applied with a camel's hair brush. A powerful discutient. Diluted it is useful over enlarged glands, for which purpose the Linimentum Iodi of the Ph. B. is also used.

Oil of Amber.

184. ℞ Sp. Camphor.. ℥ ss.
    Tinct. Opii,
    Ol. Succini....................................āā ℥ ij.
    Ol. Amygdal.. ℥ ss.
    Ft. applicatio.
To be rubbed on the chest in pertussis, &c.

Croton Oil.

185. ℞ Ol. Crotonis Tiglii.............................. ♏xx.
    Ol. Olivæ.. ℥ iij.
A powerful counter-irritant and rubefacient, producing, if much used, a pustular eruption.

22. GARGLES, THROAT APPLICATIONS, AND INHALATIONS.

Young children cannot gargle; for them, therefore, the throat should be syringed out with the gargle. Poisonous gargles should never be prescribed for children of any age—such as Belladonna, Corrosive Sublimate, and the like.

Borax and Myrrh.

186. ℞ Sodæ Bibor..................................... ʒj.
     Tinct. Myrrhæ................................... ʒss.
     Decoct Cinchonæ................................ ʒviij.
     Ft. Gargarisma.

187. ℞ Potassæ Chloratis............................. ʒj.
     Tinct. Kino..................................... ʒss.
     Aquæ.. ʒviij.
     Ft. Gargarm.

In ulceration of the fauces, relaxed throat, &c.

188. ℞ Sodæ Bibor..................................... ʒss.
     Glycerini....................................... ʒj.
     Ft. applicatio.

To be applied by means of a camel's hair brush to aphthæ, &c.

189. ℞ Sodæ Sulphitis................................. ʒj.
     Aquæ.. ʒj.

An application to be similarly used for aphthæ.

The glycerines of Carbolic and Tannic Acids of the Ph. B. are very useful applications in diseased conditions of the throat and tonsils. The glycerine of Carbolic Acid should be a little diluted for young children, and applied with care. It has been used with much success in diphtheritic exudations, foul ulcerations, &c.

Inhalations may be made by adding to half a pint of boiling water—

20 drops of Creosote (useful in ozæna and ulcerations about the pharynx attended with offensive smell), or
15 drops of Tincture of Iodine, or
20 drops of Ammoniated Tincture of Guaiacum, or
2 fluid drachms of Oil of Turpentine.

These are used more especially in chronic diseases; as, for instance, the Iodide in laryngeal phthisis. But it is also of use in severe coryza, ozæna, and some cases of bronchitis. Turpentine is an excellent stimulating vapor in cases of chronic bronchitis with much secretion. The Guaiacum is employed in throat affections, as tonsillitis, &c. Very young children, of course, cannot use inhalers. Of instruments used for inhaling, Dr. Nelson's is cheap and serviceable; there is now a great variety of instruments used for the purpose. Besides these, the following drugs are now employed in the form of atomized fluids. Hand-ball atomizers are used to produce the spray—that of Dr. Siegle is a particularly good form of instrument. Medicated spray is valuable in diseases of the mouth, fauces, pharynx, &c. A solution of Sulphurous Acid thus employed is most efficacious in sore throats, ulcerations about the tonsils, diphtheritic exudations, &c. I have employed it with marked benefit.

The following quantities of drugs are proportioned to the fluid ounce of water. Many other substances may be similarly used.

Acidi Tannici............................ 3 grains.
Aluminis Exsicc......................... 5 grains.
Boracis................................. 5 grains.
Belladonnæ Extract...................... ¼ grain.
Potass. Chlorat......................... 5 grains.
Tinct. Ferri Perchlor................... 5 drops.
Tinct. Iodi............................. 1 drop.

Atomized fluids are to be used cautiously, especially when drugs like Iodine, Opium, Belladonna, &c., are employed.

23. Liniments and Lotions.

Lotions.

Black Wash.

100. ℞ Calomelanos.. ʒj.
    Mucilag. Acaciæ...................................... ℥ss.
    Aq. Calcis.. ℥viiss.

Yellow Wash.

191. ℞ Hydrarg. Perchlorid................................. gr. vj.
    Aq. Calcis.. ℥vj.

Red Wash.

192. ℞ Zinci Sulphatis...................................... gr. vj.
    Tinct. Lavand. co.................................... ʒj.
    Aquæ... ℥vj.

Belladonna Lotion.

193. ℞ Ext. Belladonnæ..................................... Ðij.
    Aquæ... ℥viij.

Opiate Lotion.

194. ℞ Pulv. Opii.. ʒss.
    Aquæ Fervent.. ℥viij.
    Macera per horas duas et cola.

Poppy and Borax.

195. ℞ Ext. Papav... ʒij.
    Boracis.. ʒj.
    Aquæ Gervent.. ℥iv.

For itching eruptions.

Spirit.

196. ℞ Sp. Vin. Rect....................................... ℥ss.
    Aquæ Coloniensis.................................... ℥ss.
    Aquæ... ℥xv.

Cooling lotion.

Nitric Acid.

197. ℞ Acidi Nitrici dil.................................... ʒij.
    Aquæ... ℥xvj.

Alkaline Lotion.

198. ℞ Liq. Potassæ.. ʒij.
    Acid. Hydrocyan. dil................................. ʒj.
    Mist. Amygdal....................................... ℥viiss.

For itching eruptions.

Borax and Glycerine.

199. ℞ Boracis.. ʒj.
    Glycerini.. ʒj.
    Aq. Flor. Aurant. vel Rosæ......................... ʒviij.
Soothing.

Arnica.

200. ℞ Tinct. Arnicæ...................................... ʒij.
    Aquæ Sambuci...................................... ʒiv.
For sprains and bruises.

Sal Ammoniac.

201. ℞ Ammon. Hydrochlor................................. ʒss.
    Acidi Acetici...................................... ʒiiss.
    Sp. Vin. Rect.,
    Aquæ..āā ʒiij.
Discutient.

Lead.

202. ℞ Liq. Plumbi Diacetat............................... ʒj.
    Aquæ.. Oj.
Cooling.

203. ℞ Plumbi Acetat...................................... ʒj.
    Ammon. Carb:...................................... ʒj.
    Tinct. Opii....................................... ʒss.
    Aquæ Rosæ... ʒviij.
In urticaria, &c.

Carbolic Acid.

204. ℞ Acidi Carbolici.................................... ʒij.
    Sp. Rosmarini..................................... ʒj.
    Sp. Vin. Rect..................................... ʒss.
    Aquæ.. ʒvj.
For pediculi.

205. ℞ Hydrarg. Corros. Sublim........................... gr. xij.
    Sp. Vin. Rect..................................... ʒj.
    Aquæ Destillatæ................................... ʒvj.
    Ol. Rosæ.. ♏iij.
For pediculi.

Liniments.

206. ℞ Lin. Camph. co.................................... ʒj.
    Lin. Sapon. co.................................... ʒiss.
    Tinct. Opii....................................... ʒss.
For chronic pains, &c.

Opiate.

207. ℞ Tinct. Opii....................................... ʒss.
    Ext. Belladonnæ................................... Ðj.
    Lin. Sapon. co.................................... ʒiss.

Glycerine and Belladonna.

208. ℞ Ext. Belladonnæ................................... gr. xx.
    Glycerini... ʒss.

Chloroform, &c.

209. ℞ Chloroformyli .... ℨj.
     Ext. Belladonnæ. .... gr. xx.
     Ext. Opii. .... gr. xx.
     Ext. Aconiti. .... gr. x.
     Glycerini. .... ℨj.

A very powerful anodyne, to be used cautiously.

Capsicum.

210. ℞ Tinct. Capsici .... ℥ss
     Lin. Saponis co. .... ℥ss

Stimulant and Rubefacient.

Corrosive Sublimate.

211. ℞ Hydrarg. Perchlor. .... gr. ij.
     Aquæ Rosæ. .... ℥ij.
     Sp. Vin. Rect. .... ℥ss.

In favus.

212. ℞ Tinct. Cantharid. .... ℨiij.
     Lin. Sapon. co. .... ℨix.

For chilblains.

Or,

213. ℞ Calcii Chlorid. .... ℨj
     Boracis. .... ℨj.
     Axungiæ .... ℥j.

Caustic Application.

214. ℞ Acidi Chromici. .... ℨj.
     Aquæ .... ℥j.

In ringworm.

24. COLLYRIA.

Zinci Sulph., gr. j.,
Aluminis, gr. j.—viij.,
Argent. Nitrat., gr. j.—iv.;
Zinci Acetat., gr. j.,
Liq. Plumbi Diacetat., ℳv.,
Sp. Vin. Gallici, ℨj.,
} Aquæ destillatæ, ℥j.

The above are the ordinary strengths employed to the ounce of water

Zinc and Opium.

215. ℞ Zinci Sulphat. .... gr. vj.
     Vin. Opii. .... ℨj.
     Aquæ Rosæ. .... ℥vj.
     Ft. Collyrium.

To dilate the Pupil.

216. ℞ Ext. Belladonnæ. .... ℈j.
     Aquæ. .... ℥j.

Or better,

217. ℞   Atropiæ Sulphatis................................. gr. j.
          Aquæ Destillatæ.................................. ℥j.

25. Ear Lotions.

218. ℞   Calcii Chlorid.................................... ℨij.
          Aquæ... Oss.
     Ft. lotio.

219. ℞   Argent. Nitrat.................................... gr. v.
          Aquæ... ℥j.
     Ft. lotio.

220. ℞   Plumbi Acetatis,
          Zinci Sulphatis ....āā gr. x.
          Creasoti... ℔j.
     Tero simul.

The powder to be dissolved in half a pint of water. Useful in otorrhœa and offensive discharges.

26. Ointments.

Aconite.

221. ℞   Aconitiæ Puræ..................................... gr. ij.
          Cerat. Cetacei................................... ℥j.
     Misce accuratissime.
" To be used with care."

Bismuth.

222. ℞   Bismuthi Subnit................................... ℨj.
          Axungiæ.. ℨiij.
     Ft. unguent.
For cracks, excoriations, and irritable sores

Resin and Creosote.

223. ℞   Creosoti.. ℔xx.
          Ung. Resinæ,
          Adipis.......................................āā ℥j.
     Ft. unguent.
Stimulant and disinfectant.

224. ℞   Balsam. Peruv..................................... ℨij.
          Ung. Cetacei..................................... ℥j.
     Ft. unguent.
Stimulant.

Carron Oil.

225. ℞   Ol. Lini,
          Aq. Calcis, partes æquales.
A useful application for burns.

226. ℞   Collodii .... ℨij.
          Ol. Ricini....................................... ℨj.
     Misce.
An application in burns and wounds.

Carbolic Acid.

227. ℞ Acidi Carbolici... ʒj.
    Glycerini... ʒj.
    Misce.

As a dressing to wounds.

228. ℞ Ovorum Vitell.. ʒiv.
    Glycerini... ʒv.
    Misce.

An excellent coating in erysipelas, itching eruptions, &c.

Oil of Cade.

229. ℞ Ol. Cadini,
    Sulph. Præcip...................................āā ʒiij.
    Glycerini. Amyli................................ ʒvj.
    Adipis Benzoati...............................ad ʒiij.
    Ft. unguent.

Recommended by Dr. Anderson in scabies.

Iodine and Cod-liver Oil.

230. ℞ Ung. Iodi,
    Ol. Morrhuæ.....................................āā ʒiv.
    Ft. unguent.

Recommended by Dr. Tanner in bronchocele and mesenteric disease.

Calomel.

231. ℞ Calomelanos.. ʒij.
    Ung. Cetacei.. ʒj.
    Ft. unguent.

In cutaneous affections.

Scott's Ointment.

232. ℞ Ung. Hydrarg.,
    Cerat. Sapon.................................āā ʒj.
    Camph. Pulv................................... ʒj.
    Ft. unguent.

27. HYPODERMIC INJECTIONS

are eminently serviceable for the immediate relief of pain and in cases where it is desirable to get the system as rapidly as possible under the influence of the drug to be employed.

Morphia, Atropine, and Aconitine are three most potent agents when thus employed. It is wisest to commence with an extremely small dose, as the action is occasionally out of proportion to the quantity used.

Morphia. A solution of the acetate, twenty-four drops of which contain one grain of Acetate of Morphia is a convenient strength. Of this, one drop is sufficient for a child ten years old; the quantity may be carefully increased to two or three drops. The solution should be neutral.

Atropia. The Liquor Atropiæ Sulphatis (Ph. B.), each drachm of which contains half a grain of Atropine, is too strong for hypodermic use. Three fluid drachms of water should be added to one fluid drachm of the liquor; and of this solution one drop will be sufficient to commence with in a child ten years old.

Aconitine.

233. ℞ Aconitiæ... gr. j.
    Sp. Vin. Rect... ♏xv.
    Aquæ Destill... ℥j.
    Ft. solutio.
Of this solution one drop will also be sufficient at the age of ten.

CHAPTER X.

DIETARY.

1. Good Nutritious Beef Tea.

Mince one pound of good beef (from which all skin, fat, &c., have been carefully removed), and pour upon it in an earthen jar one pint of cold water. Stir, and let stand for one hour. Then place the jar in a moderate oven for one hour, or stand the jar in a saucepan of water and allow the water to boil gently for an hour. To be exact the heat to which the beef tea is raised should not exceed 180° F. Strain through a coarse sieve and allow it to go cold. When wanted remove every particle of fat from the top; warm up as much as may be required, adding a little salt. Beef tea should, except in the hottest weather, be made a day before it is wanted.

2. Essence of Beef.

One pound of gravy beef free from skin and fat, chop as fine as mincemeat, pound in a mortar with three tablespoonfuls of soft water and soak for two hours. Then put in a covered earthen jar with a little salt, cement the edges of the cover with pudding paste, and tie a piece of cloth over the top. Place the jar in a pot half full of boiling water, and keep the pot on the fire for four hours, simmering. Strain off the liquid essence through a coarse sieve: it will be about five or six ounces in quantity. One teaspoonful frequently with or without wine or brandy as may be ordered. A teaspoonful of cream may occasionally be added with advantage to four ounces of the essence, or it may be thickened with flour, arrowroot, or sago.

3. Beef Tea in Haste.

Scrape one pound of lean beef into fibres on a board. Place the scraped meat in a delicately clean white-lined saucepan and pour half a pint of boiling water upon it. Cover closely and set by the side of the fire for ten minutes, strain into a teacup, place the teacup in a basin of ice-cold water, then remove all fat from the surface, pour into a waremd cup, warm this gently with hot water or otherwise and serve. This can be ready in fifteen minutes, and double the quantity of meat can be used if necessary. Bread and blotting paper are ineffectual to remove all the fat. A tomato makes excellent flavoring, and other flavors can be added if desired. For children however, the simpler aliments are the better.

4. Beef and Chicken Broth.

One pound of good lean beef and a chicken boned should be pounded together in a mortar, a little salt added, and the whole placed in a saucepan with nearly three pints of cold water. Stir over the fire until it boils, then boil half an hour, strain through a coarse sieve, and serve.

5. Raw Meat.

Lean meat (beef, fowl, or mutton) minced finely, or grated, one part, and pure white sugar two parts, thoroughly mixed in a mortar. One teaspoonful every two, three, or four hours in diarrhœa, &c.

6. Liebig's Food for Infants.

Wheaten flour half an ounce, malt flour half an ounce, bicarbonate of potash seven and a quarter grains, water one ounce. Mix: add five ounces of cow's milk, and put the whole on a gentle fire and stir; when it begins to thicken it is to be removed; stirred five minutes; heated and stirred again till it becomes fluid, and finally made to boil; strain through a muslin sieve. Stated to be slightly aperient, and where there is tendency to diarrhœa twenty grains of prepared chalk are to be substituted for the potash.

"Laputa never devised anything more preposterous than Liebig's food for infants." Dr. King Chambers makes this remark; and despite the praise of other high authorities, I agree with Dr. Chambers. Nevertheless, let those use it who admire it.

7. Chicken, Veal, and Mutton Broths.

The fleshy part of the knuckle of veal, a chicken, bones and all, chopped up; or two pounds of the scrag end of neck of mutton, added to two pints of water with a little pepper and salt, and boiled two hours and strained, all make excellent broth. Pearl barley, rice, or vermicelli boiled separately till quite soft, may be added when either of the broths is heated for use. All fat must be always carefully removed by skimming when cold.

8. Milk and Gelatine or Isinglass.

Half an ounce of gelatine to be dissolved in half a pint of hot barley water; an ounce of powdered loaf sugar added, and a pint of new cow's milk poured in, makes an imitation of asses' milk.

Dissolve a little isinglass in water, mix well with half a pint of new milk, boil, and add sugar or not as desired.

9. Milk and Suet.

Chop one ounce of calves' suet very fine, tie lightly in a muslin bag, and boil slowly in a quart of new milk; sweeten with pounded loaf sugar. This is an imitation of goat's milk.

Boil one ounce finely chopped suet with a quarter of a pint of water for ten minutes, and press through linen or flannel. Then add one drachm of bruised cinnamon, one ounce of sugar, and three quarters of a pint of milk. Boil for ten minutes and strain. Not more than a wineglassful should be given at a time, as it is liable to derange the stomach and cause diarrhœa; a little old brandy, or a teaspoonful of La Grande Chartreuse, will prevent this, where in older children the highly nutritious and fattening qualities of this combination are desirable.

10. Milk and Lime Water.

Half a teaspoonful of the sweetened solution of lime, or an ounce to two ounces of plain lime water may be added to four ounces of new milk, or equal parts of milk and soda water make a good drink in sickness, and irritable and sour stomachs.

Fifteen grains of bicarbonate of soda added to a quart of fresh milk will prevent its turning sour for several hours, and will rather aid than impair its digestibility.

11. Bread Jelly.

Steep stale bread in boiling water and pass through a fine sieve while hot. It may be flavored and taken alone, or mixed and boiled with milk.

12. Rice Cream.

A quarter of a pound of whole rice well boiled in milk, and put in a sieve to drain and cool; mix with the rice a gill of good cream whisked to a froth, and add a wineglassful of Madeira and a little powdered loaf sugar.

13. Rice Milk.

Three tablespoonfuls of rice, one quart of milk, wash the rice and put into a saucepan with the milk, simmer till the rice is tender, stirring now and then, and sweeten. Tapioca, semolina, vermicelli, and macaroni, may be similarly prepared.

14. Rice Water.

One ounce of well-washed Carolina rice. Macerate for three hours at a gentle heat in a quart of water, and then boil slowly for an hour and strain. It may be sweetened and flavored with a little lemon peel. Useful in diarrhœa, &c., when the flavoring is best dispensed with, and a little old Cognac added.

15. Barley Water.

Wash two ounces of pearl barley with cold water, then boil for five minutes in some fresh water and throw both waters away. Then pour on a pint and a half of boiling water and boil down one-half. Flavor with thinly cut lemon rind and add sugar to taste. A little isinglass may be added if desired.

16. Rice Gruel (for Diarrhœa).

Ground rice two ounces, cinnamon a quarter of an ounce, water four pints. Boil for forty minutes, and add a tablespoonful of orange marmalade.

17. Lemonade.

The rind of three lemons pared as thin as possible should be added to a quart of boiling water, and a quarter or half an ounce of isinglass. They should stand twenty-four hours covered, then squeeze the juice of eight lemons upon half a pound of lump sugar; when the sugar is dissolved pour the lemon and water upon it, mix, strain, and serve.

18. Refreshing Drinks.

Orange, lemon, limes, or pineapple sliced small and put into a jug with an ounce or so of sugar candy. Some of the fresh juice of the fruit should be then squeezed into the jug, and a pint of boiling water poured on.

19. Tamarind Whey.

Two tablespoonfuls of tamarinds stirred into a pint of boiling milk, and strained. A quarter of an ounce of cream of tartar may be similarly treated, and a little sugar candy added.

20. Orgeat.

Two ounces of sweet almonds blanched, and a few drops of bitter almond flavor. Pound with a little orange flower water into a paste, and rub up with a pint of milk and a pint of water, until an emulsion is formed. Strain and sweeten.

21. Egg Soup.

The yolks of two eggs, a pint of water, half an ounce or so of butter and sugar to taste, beat up together over a slow fire, adding the water gradually. When it begins to boil pour backwards and forwards between the jug and saucepan till quite smooth and frothy.

22. Rose Tea.

Take of red rose buds (the white heels being removed) half an ounce, three tablespoonfuls of white wine vinegar, sugar, or sugar candy one ounce. Put into one quart of boiling water, and let stand near a fire for two hours and strain.

Similar drinks may be made with guava jelly; damson jelly; syrup of German cherry-juice; apple jelly; cape gooseberry jam, &c.

23. Jelly Water.

A dessertspoonful of wild cherry or blackberry jelly; one goblet of ice-water. Beat up well. Excellent in fever as a drink.

24. Iceland Moss Jelly.

One handful of Iceland moss well washed; one quart of boiling water; the juice of two lemons; one glass of wine; one quarter of a teaspoonful of cinnamon. The moss should be soaked an hour in a little cold water, then stirred into the boiling water, and simmer till dissolved. Sweeten, flavor, and strain into moulds; this jelly is very nourishing and is specially useful in chronic colds.

25. Isinglass Jelly.

Isinglass one ounce; pure gum arabic half an ounce; white sugar candy one ounce; port wine half a pint; a little nutmeg grated. These should be put in a jar to stand twelve hours, covered well to prevent evaporation, then placed in a saucepan with sufficient water to simmer till the contents are melted; the whole should be stirred, then allowed to stand to cool. A teaspoonful is reviving in cases of extreme exhaustion.

26. Chicken Jelly.

Half a raw chicken pounded with a mallet, bones and meat together. Cold water to cover it well. Heat slowly in a covered vessel and let it simmer until the meat is in white rags, and the liquid reduced one half; strain, and press through a coarse cloth, add a little salt, return to the fire and simmer five minutes longer, skim when cool. Wine or seasoning may be added with the salt if desirable.

27. Arrowroot Wine Jelly.

One cup of boiling water, two teaspoonfuls of arrowroot, two teaspoonfuls of white sugar, one dessertspoonful of brandy or three of wine; wet the arrowroot in a little cold water and rub smooth, then stir into the hot water, which should be on the fire and boiling at the time with the sugar already melted in it. Stir until clear, boiling all the time, and add the wine or brandy. Wet a cup in cold water and pour the jelly in to form.

A teaspoonful of lemon juice may replace the wine or brandy. The jelly can then be eaten with sugar and cream.

28. Iceland Moss and Irish Moss Jellies.

Take of Iceland moss and of Irish moss one ounce each. Boil in a pint and a half of milk slowly for three quarters of an hour. Strain through muslin and add sugar candy to taste, or an ounce of tincture of quinine may be added when more sugar will be required. One or two teaspoonfuls may be taken often in the day

29. Palatable Castor Oil.

℞ Pulv. Gum. Acac... ℥j.
    Syrupi,
    Glycerini...āā ℥j.
    Aquæ... ℥iij.
    Ol. Ricini.. ℥vj.
    Ext. Vanillæ,
    Sp. Vin. Gallic..āā ℈ij.
    Ol. Cinnam. Ver....................................... ♏v.
Misce.
Dose—Double the quantity of oil intended to be given.

30. Nutritious Enemata.—Beef Tea and Brandy.

Take of strong beef tea six ounces, one ounce of cream, half an ounce or less of brandy or an ounce of port wine. This will be sufficient for three enemata; they should be given about every eight hours unless otherwise ordered. If no other nourishment is given they will require to be given every four hours, and the stimulant should be reduced, and a few drops (say three or four) of laudanum added to control irritability of the bowel. A better plan is to chop finely the pancreas of a bullock freed from fat, and mix with eight or nine ounces of glycerine. About a fourth part of this (or less for young children) is to be added to one or more ounces of finely chopped meat and injected into the rectum as soon as made. The rectum should be cleansed with a free injection of warm water from time to time during a course of feeding, thereby to prevent irritation by decomposition of unabsorbed matters.

Quinine, cod-liver oil, bark, and other remedies can if desirable be added to nutrient enemata.

31. Stimulants.

Regarding the use of alcoholic stimulants for children I can only say that in health the less the better, and that even in disease their use is to be guarded and strictly medicinal. "Sipping from papa's glass" is a foolish and even dangerous custom, and may lay the foundation of craving for their immoderate use. Alcohol is accredited by Dr. Walshe and others with delaying the development of phthisis; that indeed "it excludes the formation of tubercle." Dr. King Chambers considers that it is rather that "the tubercles do not so soon break down into suppuration."

Whether with the object of preventing tuberculosis, or in great exhaustion, or in protracted illnesses, &c., if stimulants are to be given, what forms are best for children? As a daily drink I regard a light bitter ale, or a little good sound porter as among the most wholesome. The child should be instructed to drink rather towards the close of its meal than near the commencement.

Many of the light Hungarian, French, Greek, and Australian wines are pleasant and harmless beverages. Diluted with water, they refresh in hot weather and may assist feeble digestion. Champagne (but it must be excellent in quality) is *the* wine for a sick stomach; whether the cause be sea sickness or what not, with a small lump of ice in it, we have few more efficient remedies. Champagne is light, diffusible, easily absorbed, transitory in its effects. It is admirable where a rapid volatile stimulant is required. A few drops of old Cognac may be added in extreme prostration. Burgundy, especially the better kinds, such as Romanée, Chambertin, &c., are magnificent restorative stimulants. I have known patients recovering from exhausting illnesses remark that their glass of Burgundy seemed to "give them life." Port, if old and genuine, has also undoubted high value as a blood restorer. But the absolute necessity in sickness of having really fine wine deters one from running the risk of fusel oil and logwood. At any rate, that restorative wine is best the purity of which can best be guaranteed is a useful rule to bear in mind. I may enumerate a few very high class restorative wines, special cases, of course, indicating some, rather than others—Chateau d'Yquem, Madeira, Ruster, Red Kephesia, Como, Oberingelheimer, Steinberger cabinet, Carlowitz, Tokay, &c.

I have seen good results from the old-fashioned plan of allowing delicate young persons a glass of rum and milk early in the morning, say at least an hour, but better two hours before breakfast. The rum should be old Jamaica and a small quantity is enough. Brandy when necessary in sickness *must* be old. "Three-star Hennessy" is reliable, but whatever the kind selected may be the *older* the better, and it should be obtained where reliance can be placed on the vendor. The young, raw, fiery brandies sold are bad enough for strong stomachs; they are simply poison to the sick child. I have often, when an out-patients' physician, in crowded London districts, shuddered to hear of the "drop of brandy" and the "drop of gin" which some unhappy little one had been compelled to swallow to "do it good."

Regarding ginger and orange and the other "home-made" wines they are innocent enough, except that with some children they are apt to produce biliousness, or to turn sour on the stomach.

INDEX.

Entry	Page
Abdominal tumors	155
Acids	168
Acne	35
Aconite	182
— treatment by	112
Active congestion	71
Acute desquamative nephritis	158
— hydrocephalus	74
— laryngitis	109
— peritonitis	152
Ague	58
Alcohol	176
Albuminoid liver	155
Alteratives	169
Antacids	164
Anthelmintics	150, 190
Antiseptics	177
Antispasmodics	175
Aphorisms of Bouchut	4
Apomorphia	187
Apoplexy, cerebral	73
— meningeal	74
Arrowroot jelly	208
Ascarides	150
Asphyxia neonatorum	39
Astringents	167
Atelectasis pulmonum	121
Barley water	206
Bath, cooled	50, 55
Baths	194
Beef tea	204
Blood-restorers	162
Brain, congestion of	71
Bread jelly	206
Bronchitis	110
Broncho-pneumonia	111
Bullæ	31
Cancer of the liver	157
— of the kidney	157
— of the stomach	157
Cancrum oris	136
Catarrhal pharyngitis	137
Cephalhæmatoma	93
Chicken broth	205
— jelly	208
Chicken-pox	64
Child crowing	116
Chloasma	37
Chorea	86
Chronic hydrocephalus	82
Chronic peritonitis	153
Cod-liver oil	164
Colic	143
Collyria	200
Congenital diseases	39
Congestion of the brain	71
Contraction with rigidity	70
— essential	70
— symptomatic	70
Convulsions	67
Cooled bath	50, 55
Coryza	97
Counter-irritants	196
Cow-pox	63
Croup	102
— diagnosis of	108
— spasmodic	107
— spurious	107
Cyanosis	128
Cynanche parotidea	138
— trachealis	102
Dentition	10
Diabetes	157
Diaphoretics	192
Diarrhœa	144
— chronic	145
— inflammatory	148
Diet tables	12
Dietary	204
Digitalis in heart disease	131
Diphtheria	98
Diuresis	157
Diuretics	191
Dysentery	148
Dyspepsia	140
Dysuria	159
Ear lotions	201
Eclampsia nutans	89
Ecthyma	33
Eczema	29
Effusion, pericardial	130
Egg soup	207
Emetics	187
Encephalitis	80
Endocarditis	129
Enteralgia	143
Epilepsy	89
Epistaxis	133
Erysipelas	28
Erythema	27

INDEX.

	PAGE
Erythema nodosum	27
Essence of beef	204
Exanthemata	27
Expectorants	193
Expression, significance of	2
External applications	194
Eye, condition of	4
Fit, immediate treatment of	68
Food, varieties of	6
Formulary	162
Gargles	196
Gastritis	142
Gastric catarrh	142
Gelsemin	174
General indications	1
— therapeutic hints	161
Gestures, significance of	3
Hæmorrhage, cerebral	73
Herpes	30
— circinatus	31
— zoster	31
Hooping-cough	113
Hydatids of the liver	156
Hydrocephalus, acute	74
— chronic	82
Hydrocephaloid disease	79
Hydrorachis	95
Hypertrophy of the brain	81
— of the heart	130
Hypodermic injections	202
Iceland Moss jelly	208
Icterus neonatorum	152
Icthyosis	35
Idiocy	66
Impetigo	32
— figurata	32
Incontinence of urine	158
Infantile remittent fever	51
Inhalations	196
Intelligence, marks of	66
Intermittent fever	58
Internal disinfecting	49
Intertrigo	27
Inward fits	68
Irish moss jelly	209
Isinglass jelly	208
Jaundice	152
Jelly water	208
Laryngismus stridulus	110
Laryngitis	109
Laxatives	188
Lemonade	207
Lepra	34
Leucocythæmia	157
Lichen	33
— agrius	34
— strophulus	33
— urticatus	33

	PAGE
Liebig's food	205
Light, value of	10
Liniments	198
Liver, fatty	155
— albuminoid	155
— amyloid	155
— cancer of	157
— hydatids of	156
Lotions	198
Lupus	35
Management during first year	5
Measles	43
Miliaria	29
Milk and gelatine	205
— and lime water	206
— and suet	205
Milks, composition of	9
Molluscum	35
Mumps	138
Mutton broth	205
Navel, diseases of	5, 39
Nephritis, acute desquamative	158
Nervine tonics	183
Nettle rash	28
New-born, diseases of the	39
Night terrors	71
Noma	135
Nutritious enemata	209
Ointments	201
Ophthalmia, strumous	15
— neonatorum	41
Operation of tracheotomy	106
Orgeat	207
Otorrhœa	92
— strumous	15
Ozæna	15
Palatable castor oil	209
Papulæ	33
Paracentesis thoracis	124
Paralysis	90
— atrophic infantile	91
— facial	92
— progressive myosclerotic	91
Parasitici	36
Parotitis	138
Passive congestion	73
Pemphigus	31
Pericarditis	129
Peritonitis, acute	152
— tubercular	153
Pertussis	113
Phosphorus	176
Phthisis	125
Pitting, to prevent	63
Pityriasis	35
Pleurisy	122
Pneumonia	117
— lobar and lobular	119
Pompholyx	31
Prolapsus ani	160

INDEX.

	PAGE
Prurigo	34
Psoriasis	34
Pulse, table of	4
Purgatives	189
— anthelmintic	190
— cholagogue	189
— drastic	189
— hydragogue	189
— saline	189
Pustulæ	32
Quinsy	138
Rachitis	18
Raw meat	205
Refreshing drinks	207
Retention of urine	159
Retropharyngeal abscess	140
Respiration	2
Rheumatism	24
Rice cream	206
— milk	206
— gruel	207
— water	206
Rickets	18
Ringworm	36
Rose tea	207
Roseola	27
Rötheln	45
Rubeola	43
— notha	45
Rules for prescribing	161
Rupia	32
Salaam convulsion	89
Salicin	25, 185
Scabies	37
Scarlatina	45
— anginosa	47
— latens	47
— maligna	47
Scarlatinal bubo	47
Sclerema	40
Scrofulosis	14
Scutellarin	174
Sedatives to the brain	178
— cardiac	180
— general	178
— to the spinal cord	174
— vascular	180
Smallpox	59
Sore throat	137
Spastic rigidity	70
Spina bifida	95
Spinal cord, irritation of	93
— chronic inflammation	94
— inflammation of	93
Spurious hydrocephalus	79
Squamæ	34
St. Vitus's dance	86
Stimulants	175, 209
— ganglionic	175
— in typhus and typhoid	57
— to the brain	175

	PAGE
Stimulants to the spinal cord	172
— vascular	175
Stomachics	186
Stomatitis	135
— gangrenous	136
— ulcerative	135
Struma	14
Strumous abscess	14
— ophthalmia	15
— otorrhœa	15
— ozæna	15
Sudamina	29
Syphilis	21
Tabes mesenterica	154
Table of Gaubius	162
Tamarind whey	207
Teeth, order of appearance	10
Temperature	3
Tetanus neonatorum	94
Thrush	134
Tinea tonsurans	36
— decalvans	87
— favosa	36
Tongue, condition of the	3
Tonsillitis	138
Tracheotomy	106
Trismus	94
Tubercle in the brain	84
Tubercula	35
Tubercular meningitis	74
— peritonitis	153
Tuberculosis	16
Tumours, abdominal	155
Typhoid fever	51
Typhus fever	56
Typhus fever and typhoid diagnostic table	57
Ulcerations in the throat	137
Urine, incontinence of	158
— retention of	159
Urticaria	28
Vaccination	63
Vaccinia	63
Vaginitis	160
— scarlatinal	48
Varicella	64
Variola	59
— confluens	60
— discreta	59
— nigra	60
Varioloid	61
Veal broth	205
Veratrum viride, treatment by	112
Vesiculæ	29
Vomiting	8
Weaning	9
Wet-nurse, choice of	8
Wet sheet	50
Worms	149

www.ingramcontent.com/pod-product-compliance
Lightning Source LLC
Chambersburg PA
CBHW021831230426
43669CB00008B/939